# SOLVENT PROBLEMS IN INDUSTRY

This volume contains a collection of papers presented at the 3rd and 4th European Solvents Symposia for Industry, held at the University of Manchester Institute of Science and Technology, Manchester, UK.

# SOLVENT PROBLEMS IN INDUSTRY

*Edited by*

## GEORGE KAKABADSE

Dipl. Ing., Dr Ing., C. Chem., FRSC

*Department of Chemistry,
University of Manchester Institute of Science and Technology,
Manchester, UK*

## ELSEVIER APPLIED SCIENCE PUBLISHERS
### LONDON and NEW YORK

7221-3917
1/04 TR. from

FOREST PRODUCTS

ELSEVIER APPLIED SCIENCE PUBLISHERS LTD
Ripple Road, Barking, Essex, England

*Sole Distributor in the USA and Canada*
ELSEVIER SCIENCE PUBLISHING CO., INC.
52 Vanderbilt Avenus, New York, NY 10017, USA

**British Library Cataloguing in Publication Data**

Solvent problems in industry.
1. Solvents
I. Kakabadse, George
661'.807      TP247.5

ISBN 0–85334–304–7

WITH 55 TABLES AND 64 ILLUSTRATIONS

© ELSEVIER APPLIED SCIENCE PUBLISHERS LTD 1984

Printed in Northern Ireland by The Universities Press (Belfast) Ltd.

# PREFACE

This book places emphasis on solvent problems in industry and aims to rationalise the relationship between solvent properties and economic use of solvents. It is largely a report of the proceedings of the 'European (4th UMIST) Solvents Symposium for Industry', jointly organised by industry and UMIST in 1983. The scope of this volume has been enlarged by including some material presented at the 3rd Solvents Symposium for Industry in 1980 and duly updated in 1983. The excellent attendance at the four solvents symposia for industry (1974–83), the first of their kind, is a clear indication of a need for presenting and discussing solvent problems in industry and for highlighting new developments. In this book, solvent topics are viewed under four sections, viz. 'General', dealing with fundamental and broader aspects of solvents; 'Applied', listing selected solvent applications; 'Solvent Recovery and Disposal', stressing the economic and environmental issues involved; and finally, the ever changing aspects concerning 'Health, Legislation and Safety', including the latest EEC directives for labelling of solvents.

The text is intended mainly for industrial personnel dealing with solvents but it may also prove useful to chemists, chemical engineers and management scientists at universities and polytechnics interested in solvents, which have a vast market.

I am greatly indebted to my industrial co-organisers and their respective companies without whose cooperation and assistance the solvents symposia would not have been possible: Charles Andrews (Solvents Consultant), Brian Atkinson (Petrofina UK), Giuliano Ballini (Montedipe), Brian Davis (Hoechst UK), Barry Hudson (BP Chemicals), John Revell (Ford Motor Company), Mike Smith (Esso Chemicals), Richard Williams (Shell Chemicals), Ken Wright (ICI plc, Organics Division). I should also like to express my thanks to my wife for proof-reading and constant encouragement and to the following colleagues at UMIST: John Belcher, Alan Beresford, Jenny Curtis, Roger Perry, Bob Ramage, Neville Smith and Tony Tipping.

Particular thanks are given to my co-authors, recognised experts in their own field, who willingly interrupted their busy routine to present an up-to-date account of their invited papers.

GEORGE KAKABADSE

v

# CONTENTS

# LIST OF CONTRIBUTORS

C. W. ANDREWS
*Solvents Consultant, 'Lamorna', Martineau Drive, Dorking, Surrey RH4 2PL, UK*

L. E. BAKER
*Development Director, Re-Chem International Ltd, 80 Shirley Road, Southampton SO1 3EY, UK*

A. BERLIN
*Principal Administrator, Commission of the European Communities, Plateau du Kirchberg, BP 1907, Luxembourg*

M. J. BLANDAMER
*Senior Lecturer, Department of Chemistry, University of Leicester, Leicester LE1 7RH, UK*

J. C. BODEN
*Project Leader, Combustion Research, BP Research Centre, Chertsey Road, Sunbury-on-Thames, Middlesex TW16 7LN, UK*

A. V. BRIDGWATER
*Senior Lecturer, Department of Chemical Engineering, University of Aston in Birmingham, Gosta Green, Birmingham B4 7ET, UK*

S. FUMERO
*Administrator, Commission of the European Communities, Brussels, Belgium.* Present address: *Istituto Recerche Biomedicle, 'A. Merxa' RBM, CP 226, 10015 Ivrea, Italy*

D. GEE
*National Health and Safety Officer, General and Municipal Workers' Union, Thorne House, Ruxley Ridge, Claygate, Esher, Surrey KT10 0TL, UK*

S. Hewerdine
*Materials Engineer, ICI plc, Engineering Department, Materials Group, PO Box 6, Billingham, Cleveland, TS23 1LD, UK*

G. H. Hutchinson
*Formerly Technical Director, Croda Inks Ltd, Edinburgh.* Present address: *Consultant, Croda Inks Ltd, Park Works, Park Lane, Harefield, Middlesex UB9 6HQ, UK*

G. J. Kakabadse
*Honorary Senior Lecturer, Department of Chemistry, UMIST, PO Box 88, Manchester M60 1QD, UK*

F. C. Lloyd
*Research Physicist, ICI plc, Organics Division, PO Box No 42, Hexagon House, Blackley, Manchester M9 3DA, UK*

I. Metcalfe
*Director, Sutcliffe Speakman Engineering Ltd, Guest Street, Leigh, Lancashire WN7 2HE, UK*

I. C. Miller
*Development Engineer, Ford Motor Co. Ltd, Arisdale Avenue, South Ockendon, Essex RM15 5TJ, UK*

G. Mosselmans
*Principal Administrator, DGIII, Internal Market and Industrial Affairs, Commission of the European Communities, RP3 – 4/9, Rue de la Loi 200, B-1049 Brussels, Belgium*

W. G. Murcar
*Responsible for Business Development, Kaldair Ltd, Astronaut House, Hounslow Road, Feltham, Middlesex TW14 9AD, UK*

C. Penning
*Commercial Manager, Paktank Industriele Dienstverlening BV, Onde Maasweg 6, Rotterdam, The Netherlands*

K. G. Reed
*Senior Research Associate, Solvents Technology Division,*

*Essochem Europe Inc., Nieuwe Nijverheidslaan 2, B-1920 Machelen, Belgium*

C. Reichardt
*Professor of Organic Chemistry, Philipps University, Hans Meerwein Strasse, D-3550 Marburg, Federal Republic of Germany*

J. M. C. Roberts
*Product Research Manager, Johnson Wax Ltd, Milton Park, Stroude Road, Egham, Surrey PW20 9UH, UK*

L. Turner
*Head of R & D Solvents Section, Shell Research BV, Badhuisweg 3, Amsterdam, The Netherlands*

M. J. Welch
*Senior Development Manager, Evode Ltd, Common Road, Stafford ST16 3EH, UK*

B. P. Whim
*Head of Solvents Marketing Department, ICI plc. Mond Division, PO Box 19, Runcorn, Cheshire WA7 4LW, UK*

R. S. Whitehouse
*Forward Project Manager, Evode Ltd, Common Road, Stafford ST16 3EH, UK*

C. S. H. Wilkins
*Formerly Managing Director, Sutcliffe Speakman Engineering Ltd, Guest Street, Leigh, Lancashire WN7 2HE, UK*

M. R. Wright
*HM Principal Inspector of Factories, Health and Safety Executive, 39 Baddow Road, Chelmsford, Essex CM2 OHL, UK*

# PART I

# GENERAL ASPECTS

# 1

## WHAT ARE SOLVENTS?

MICHAEL J. BLANDAMER

*Department of Chemistry, The University, Leicester, UK*

## ABSTRACT

*In an ideal solution, each solute molecule is independent of all other solute molecules. Thus solvents keep solute molecules apart. A related task for solvents is to disperse pure substances (solid or liquid) into solution. Both dispersal and separation require some stabilisation of solutes in solution through solvent–solute interactions. Solvents can also support otherwise transient species, e.g. the proton as $H_3O^+$ and the electron as the 'solvated electron'. In other systems, solvents play a key rôle in determining the conformation of solutes.*

*The position of a chemical equilibrium can be altered by appropriate choice of solvent (and the temperature of the solvent). New solute species, such as ion-pairs, can be formed in solution by choosing a solvent having low permittivity. A similar rôle is played by solvents in controlling the rate of chemical reaction. Some solvents are also chosen because they provide a medium for catalytic reactions.*

## 1. INTRODUCTION

As a matter of definition, a solution comprises at least two components.[1] The component which is in large molar excess, and normally a liquid in the pure state at the same temperature and pressure as the solution, is called the *solvent*. The minor component is called the *solute*. Somewhat perversely, when discussing the properties of solutions our main concern is with the properties of solutes. Often a particular solvent is chosen because there is a particular property of a solute which we wish to either develop or exploit. Therefore in attempting to answer the question posed by the title to this article, we consider how solvents influence the properties of solutes. This is an enormous subject. Here we use thermodynamics as the basis for our discussion.

## 2. CHEMICAL POTENTIALS

We confine attention to systems held at constant temperature and constant pressure. The dependent extensive variable called the Gibbs function can be defined by the set of independent variables $T$, $p$ and $n_i$, where $n_i$ represents the number of moles of independently defined chemical substances in the system $n_1$, $n_2$, $n_3. \ldots, n_i$. The chemical potential $\mu_j$ of a given substance $j$ (i.e. one of the set of $i$-substances) is defined by the following partial derivative:

$$\mu_j = (\partial G/\partial n_j)_{T,p,n_{i \neq j}} \tag{1}$$

If substance $j$ is a solute in solution, the chemical potential of $j$ is related, by definition,[2] to the molality $m_j$ by the equation:

$$\mu_j(\text{sln}; T; p) = \mu_j^{\ominus}(\text{sln}; T) + RT \ln (m_j \gamma_j/m^{\ominus}) + \int_{p^{\ominus}}^{p} V_j^{\infty}(\text{sln}; T; p) \, dp \tag{2}$$

Here $\mu_j^{\ominus}(\text{sln}; T)$ is the chemical potential of $j$ in its solution standard state where $m_j = 1$, $\gamma_j = 1$ at temperature $T$ and standard pressure $p^{\ominus}$. The quantity $V_j^{\infty}(\text{sln}; T; p)$ is the partial molar volume of $j$ in solution at infinite dilution. The activity coefficient $\gamma_j$ is defined using:

$$\text{Lt} \, (m_j \to 0) \gamma_j = 1 \cdot 0 \tag{3}$$

at all $T$ and $p$. Thus in an ideal solution $\gamma_j = 1 \cdot 0$; each solute molecule behaves independently of all other solute molecules in solution. In these terms, the basic rôle of a solvent is to keep solute molecules apart and thereby to insulate them from each other. In real solutions, solute–solute interactions account for deviations from ideal. Indeed these deviations can be quite marked where solute–solute interactions are strong and long-range, e.g. between ions in a salt solution.[3]

We have turned the question concerning solvents into a discussion of the properties of solutes. However the terms, solute and solvent, are simply convenient labels attached to different components of the same system. Indeed the properties of all components in a system are linked through the Gibbs–Duhem equation:

$$S\,dT - V\,dp + \sum_{j=1}^{j=i} n_j\,d\mu_j = 0 \tag{4}$$

For solutions, eqn (4) describes a communication between solutes and solvent.

## 3. STANDARD CHEMICAL POTENTIALS

The standard chemical potential $\mu_j^{\ominus}(\text{sln}; T)$ of solute $j$ in solution is the parent of a family of standard partial molar properties including the standard enthalpy $H_j^{\ominus}$, standard volume $V_j^{\ominus}$, standard entropy $S_j^{\ominus}$ and standard heat capacity at constant pressure $Cp_j^{\ominus}$. These quantities are related through well-known thermodynamic equations; e.g. the Gibbs–Helmholtz equation:

$$[\partial(\mu_j^{\ominus}/T)/\partial T] = -H_j^{\ominus}/T^2 \tag{5}$$

Nevertheless we cannot calculate an absolute value for $\mu_j^{\ominus}(\text{sln}; T)$. All that we can do is to discuss differences between the chemical potentials of $j$ in different standard states. Thus with reference to two phases, $\alpha$ and $\beta$, we can consider the quantity $\Delta(\alpha{\rightarrow}\beta)\mu_j^{\ominus}(T)$ and related parameters such as $\Delta(\alpha{\rightarrow}\beta)H_j^{\ominus}(T)$. These $\Delta$-quantities can be thought of as describing the *transfer* of $j$ from $\alpha$ to $\beta$. Here we meet another rôle played by solvents, namely *dispersal*. If, for example, $\alpha$ is the pure component $j$ (solid or liquid) and $\beta$ is a solvent, the quantity $\Delta(\alpha{\rightarrow}\beta)\mu_j^{\ominus}(T)$ refers to the dispersal of pure $j$ into solution. This transfer is characterised experimentally by the solubility. The processes of dispersal and separation rely on the stabilising influence of solute–solvent interactions.

## 4. SOLUTE–SOLVENT INTERACTIONS

An alternative heading to this section would be 'Intermolecular Forces' because we are concerned with the interaction between solute $j$ and neighbouring solvent molecules and, as important, the interactions between solvent molecules which surround solute $j$. A useful approach in this context was suggested by Gurney.[4] A solute in solution is

surrounded by a co-sphere of solvent within which the organisation of
solvent differs from that of the bulk solvent outside the co-sphere.
Hence solute–solvent interactions are described in terms of solute–
co-sphere solvent interactions. The range of these interactions is briefly
commented on in this section.

The important rôle played by van der Waal forces in solute–solvent
interactions is borne out by solubility data[5] for non-electrolytes in
liquids. These data can be considered from two standpoints[6,7] (Fig. 1).
The solubilities of a given solute in a range of solvents are related to
the Hildebrand solubility parameter $\delta_1$ $[=(\Delta_v E_1^{\ominus}/V_1^*)^{1/2}$ where $1 =$
solvent]. Thus $\delta_1$ is a measure of the strength of intermolecular forces
within a solvent. Alternatively the solubilities in a given solvent of a
number of solutes are related to the Lennard–Jones force constant for
gaseous solutes, $\varepsilon$. The latter quantity provides a measure of the
repulsion–dispersion forces involving the solute.

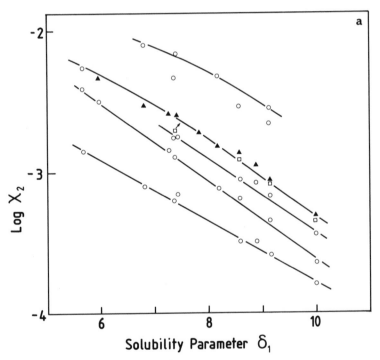

Fig. 1. Solubilities of apolar solutes: (a) dependence of solubility on
Hildebrand parameter $\delta_1$ for solvent.

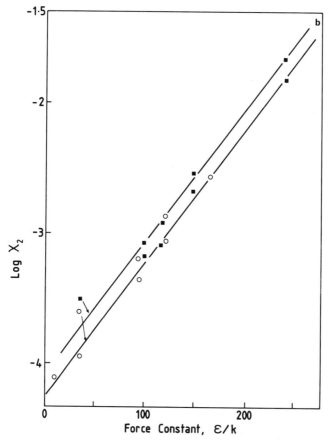

Fig. 1—*contd.* (b) Dependence of solubility in a given solvent on Lennard–Jones force constant for solute (re-drawn from Ref. 7).

The subtlety of solute–solvent interactions is borne out by the dependence of solubility of apolar solutes on the composition of a liquid mixture. In the example[8] given in Fig. 2, we see how the solubility can be controlled by modest changes in the composition of the solvent, although the underlying relationship between solubility and composition is complicated.

With increase in intensity of solute–solvent interactions, so the possibility emerges of a specific interaction between solute $j$ and a single neighbouring solvent molecule. The high solubility of iodine in

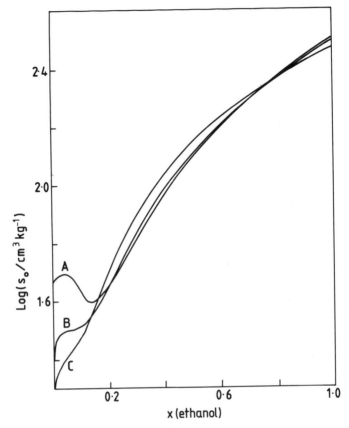

Fig. 2. Solubility of argon in ethanol + water mixtures as a function of mole fraction of alcohol, $x_2$; (a) 227·2 K, (b) 303·0 K and (c) 333·3 K (re-drawn from Ref. 8).

benzene and the intense colour of many otherwise simple solutions provides the starting point for examination of charge–transfer complexes and donor–acceptor complexes.[9]

A particularly important interaction between solute and solvent is hydrogen bonding.[10] Spectroscopic evidence for this interaction is compelling. For example,[11] very dilute solutions of $CH_3OH$ in tetrachloromethane show an absorbance band in the $3600 \text{ cm}^{-1}$ region which is assigned to the O—H stretching vibration of $CH_3OH$ (Fig. 3). When $CH_3I$ is added to the solution, the intensity of this band

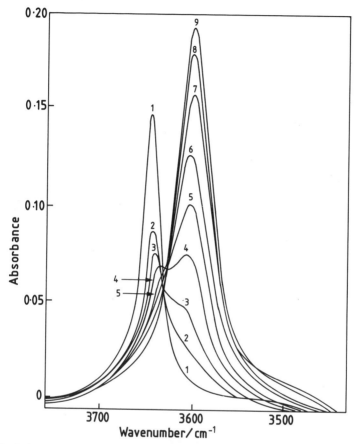

Fig. 3. Infra-red absorption spectra of a very dilute solution of methyl alcohol in tetrachloromethane; mole fraction of $CH_3I$ = (1) 0, (2) 0·13, (3) 0·24, (4) 0·44, (5) 0·61, (6) 0·76, (7) 0·89, (8) 0·94, (9) approx. 1·00 (re-drawn from Ref. 11).

decreases whereas the intensity of a new band at lower wavenumber increases. The new band is assigned to the O—H stretching vibration within the species, $CH_3.O$—$H \ldots I.CH_3$.

If the intensity of solute–solvent interaction is sufficiently strong, bond fission may occur. This process is shown diagrammatically in Fig. 4 for a solute molecule A—B. The covalent bond between groups A and B can break either homolytically or heterolytically. In the gas phase,

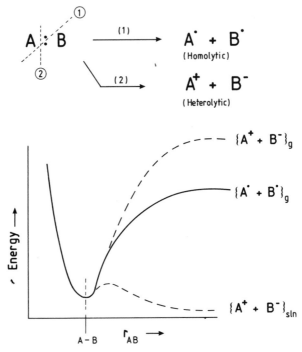

Fig. 4. Effect of solvent on energetics of bond fission for a simple diatomic molecule AB.

homolytic fission is often the preferred pathway (i.e. a lower energy barrier) but in solution, intense solvent–solute interaction can swing the balance in favour of heterolytic fission. In these terms, solvents are used to generate new solute species, e.g. $HCl(g) + H_2O(\ell) \rightarrow H_3O^+(aq.) + Cl^-(aq.)$.

Some indication of the intensity of ion–solvent interactions is obtained from standard enthalpies of solution for salts into solution; i.e. $\Delta(s \rightarrow sln)H^{\ominus}(sln; T)$. Actually these enthalpies are usually small. In other words, ion–solvent interactions for salts in solution are comparable in strength to ion–lattice interactions in the crystalline state.[12] The organisation of solvent within the co-sphere around an ion in solution remains a topic for debate. For simple ions (e.g. alkali metal cations and halide ions) in aqueous solution, the co-sphere may comprise two component parts.[13] The organisation of solvent in the outer part has been described as disordered.[14] Recent advances in neutron and x-ray scattering techniques[15] have revealed important information concerning the organisation of solvent molecules in the inner solvation shells of ions[16] (Fig. 5). Details of ion–solvent molecule interactions have also emerged from mass spectra of ion–solvent clusters.[17]

Solute–solvent interactions can be so powerful that otherwise transient species are stabilised in solution. Classical acid–base chemistry is based on the stability of the proton as $H_3O^+$ in aqueous solution.[18] The structure of the analogous $D_3O^+$ is shown[19] in Fig. 6. Similarly the solvated electron[20] $e^-$(solv.) is the simplest anion. Here electrons are trapped within a potential energy well formed by solvent molecules.[21] Support for the idea that $e^-$(solv.) is the simplest anion is given by the data shown in Fig. 7. The latter compares the energies of the low energy absorption bands of iodide ions and $e^-$(aq.) in a range of solvent systems. In the case of the iodide ion, the band is assigned to a charge-transfer-to-solvent transition[22] where the solvent plays a key rôle in defining the energy level for the excited state. Neighbouring

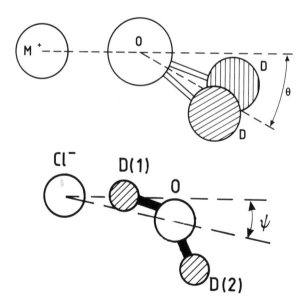

Fig. 5. Orientation of water molecules in primary solvation shell of ions in aqueous solution; LiCl in $D_2O$ where $\theta = 40 \pm 10°$ and $\psi = 0°$ where [LiCl] = $3.57\ \mathrm{mol\ dm^{-3}}$ (re-drawn from Ref. 16).

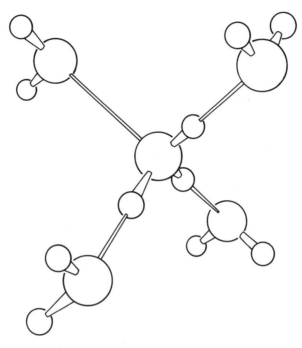

Fig. 6. Structure of $D_3O^+$ in concentrated DCl solutions in $D_2O$ (re-drawn from Ref. 19).

solvent molecules play a similar rôle in defining the energy levels for $e^-$(solv.) in the trapping cavity formed by the solvent.

## 5. CHEMICAL EQUILIBRIA

At fixed temperature and pressure, all spontaneous processes within a closed system operate to lower the Gibbs function $G$ of the system. Chemical equilibrium corresponds to a minimum in $G$ where the affinity for spontaneous change is zero. The composition of a system at the minimum in $G$ is characterised by the standard equilibrium constant $K^\circ$. By changing the solvent, this equilibrium composition can be altered. A more subtle change in solvent can be brought about by changing the temperature. We illustrate the point by reference to the data for the dependence of $K^\circ$ on temperature for acetic acid[23] in

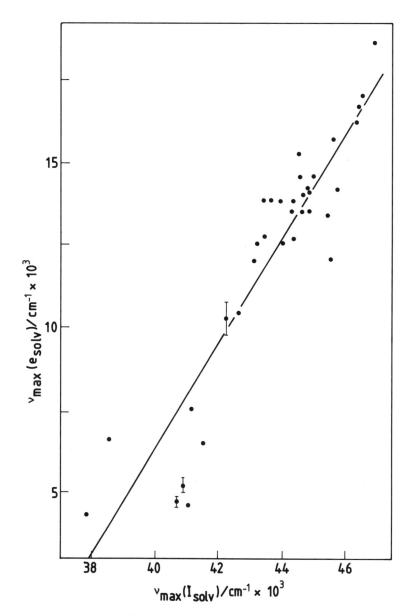

Fig. 7. Plot of $\nu_{max}(e^-)$ against $\mu_{max}(I^-)$ for 35 different solvent systems (re-drawn from Ref. 21).

MICHAEL J. BLANDAMER

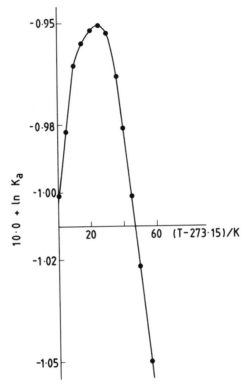

Fig. 8. Dependence on temperature of acid dissociation constant for acetic acid in water (data taken from Ref. 23).

water, Fig. 8. Thus by changing the temperature the dissociation process moves from endothermic to exothermic. Indeed at room temperature the dissociation is essentially athermal.[24] This latter feature would be difficult to predict simply on the basis of the following equation:

$$CH_3.COOH + H_2O \rightleftharpoons H_3O^+(aq.) + (CH_3COO^-)_{aq}.$$

This example highlights another feature of solvents—their moderating influence.

Solvents also play an important rôle in conformational equilibria. By changing the solvent, the preferred conformation of a polymer may change from, for example, coiled to extended form.[25] Similarly the catalytic function of enzymes crucially depends on the way in which the solvent supports the required tertiary structure.[26,27]

By changing the solvent, new solute species can be produced. If, for example, the relative electric permittivity of the solvent in a solution of a $1:1$ salt is lowered, a fraction of the free ions may associate to form ion-pairs.[28] The equilibrium between free ions and ion-pairs can be shifted to favour ion-pair formation by decreasing the relative permittivity.[28,29] Confirmatory evidence for ion-pairs in solution is obtained from electron-spin resonance[30] and ultra-violet[22] spectroscopy. In other systems, solvents can be used to induce clustering to form dimers, trimers, . . . , etc., e.g. micelle formation.

## 6. KINETICS OF REACTION

The rate of approach of a system towards equilibrium can be characterised by one or more rate constants. Rate constants for chemical reactions are particularly sensitive to solvent.[31] In other words, solvents can be used to control the rate of reaction. Indeed the control can bias the formation of a particular product at the expense of others. Qualitative guidelines[32] relate the effect of solvent on the rate of reaction to the change in electrical properties of the solutes on activation. Quantitative treatments based on measures of solvent polarity[33] (e.g. $Y$-values[34,35]) are also used. Nevertheless a prediction[36] 'a priori' of the effect of changing the solvent on the rate constant for a given reaction is not generally successful. Interpretation of trends in derived parameters (e.g. standard enthalpies of activation) poses severe problems especially where the dependence of these functions on solvent shows complicated patterns, 'roller-coaster' effects.[37] A number of problems arise in this connection stemming in part from the fact that rate constants are phenomenological quantities. Unless one is reasonably confident that a given rate constant is characteristic of one discrete process rather than of many contributory rate and equilibrium parameters, analysis of solvent effects is not a simple task.[38] Where such complexities are absent, the calculated activation parameters represent a difference between the corresponding standard partial molar properties of initial and transition states. More detailed analysis requires information from, for example, thermodynamic data concerning the effect of solvent on the initial states. Combination of these data with kinetic data can yield detailed insight into the factors controlling the rate of reaction.[39]

For open systems in far-from-equilibrium states, non-linear effects in chemical kinetics can lead to dissipative structures. The rôle of solvents in this class of reactions has not been examined to any depth. Nevertheless a key consideration here is often the ability to support both redox and acid–base controlled reactions.

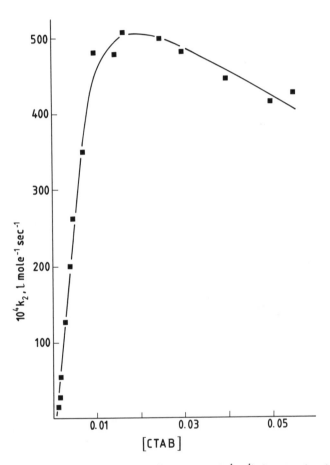

Fig. 9. Second order rate constant ($k_2$, litre $mol^{-1} s^{-1}$) for reaction between hydroxide ions and 2,4-dinitrochlorobenzene in aqueous solution as a function of concentration of CTAB (re-drawn from Ref. 40).

## 7. CATALYSIS

Solvents are used as media for catalysis reactions. The solvent carries out a number of functions. Clearly the solvent must support the catalyst and, less obviously but as importantly, it must provide a solvent for reactions and products. We commented above on enzymic catalysis. Data[40] for a micellar catalysed reaction[41] are summarised in Fig. 9 for a reaction between hydroxide ions and dinitrochlorobenzene in aqueous solution containing cetyltrimethylammonium bromide (CTAB) micelles. More complicated liquid systems can also be prepared in which the rate of chemical reaction is significantly accelerated. For example, we have reported[42] recently how certain chemical reactions in microemulsions[43] are significantly faster than in more conventional solutions. This acceleration can be understood in terms of the microheterogeneous character of these systems.[44]

## 8. CONCLUSION

We have attempted to answer the question 'What are solvents?' in operational terms. To start with we developed the idea of separation, dispersal and stabilisation. These terms led to the concepts of moderation and control by solvents in the context of chemical equilibria and chemical kinetics. We have shown how solvents exert considerable control over the chemistry of solutes. In many respects this control is their 'raison d'être'.

## REFERENCES

1. G. N. Lewis and M. Randall, *Thermodynamics*, McGraw-Hill, New York, 1923, p. 31.
2. M. L. McGlashan, *Chemical Thermodynamics*, Academic Press, London, 1979.
3. R. A. Robinson and R. H. Stokes, *Electrolyte Solutions* (2nd edn), Butterworths, London, 1959.
4. R. W. Gurney, *Ionic Processes in Solution*, McGraw-Hill, New York, 1953.
5. R. Battino and H. L. Clever, *Chem. Rev.*, 1966, **66,** 395.
6. J. H. Hildebrand and R. L. Scott, *The Solubility of Non-electrolytes* (3rd edn), Dover, New York, 1964.
7. J. E. Jolley and J. H. Hildebrand, *J. Am. Chem. Soc.*, 1958, **80,** 1050.

8. R. W. Cargill and T. J. Morrison, *J. Chem. Soc., Faraday Trans. I*, 1975, **71**, 618.
9. R. Foster, *Organic Charge-Transfer Complexes*, Academic Press, London, 1969.
10. P. A. Kollman and L. C. Allen, *Chem. Rev.*, 1972, **72**, 283.
11. M. C. R. Symons, N. G. M. Pay and G. Eaton, *J. Chem. Soc., Faraday Trans. I*, 1982, **78**, 1841.
12. B. E. Conway and J. O'M. Bockris, *Modern Aspects of Electrochemistry*, Butterworths, London, 1954, Chapter 2.
13. H. L. Friedman and C. V. Krishnan, in *Water—A Comprehensive Treatise* (ed. by F. Franks), Plenum Press, New York, 1973, Vol. 3, Chapter 1.
14. H. S. Frank and W. Y. Wen, *Disc. Faraday Soc.*, 1957, **24**, 133.
15. J. E. Enderby and G. W. Neilson, *Rep. Progr. Phys.*, 1981, **44**, 593.
16. J. E. Enderby and G. W. Neilson, reference 13, Vol. 6, Chapter 1.
17. P. Kebarle, *Modern Aspects of Electrochemistry* (ed. by B. E. Conway and J. O'M. Bockris), Plenum Press, 1974, Vol. 9, p. 1.
18. M. Eigen, *Angew. Chem. Intern. Edn*, 1964, **3**, 1.
19. R. Triolo and A. H. Narten, *J. Chem. Phys.*, 1975, **83**, 3624.
20. E. J. Hart (Ed.), *The Solvated Electron*, Advances in Chemistry Series, No. 50, American Chemical Society, Washington, DC, 1965.
21. M. C. R. Symons, *Chem. Soc. Rev.*, 1976, **5**, 337.
22. M. J. Blandamer and M. F. Fox, *Chem. Rev.*, 1970, **70**, 59.
23. H. S. Harned and R. W. Ehlers, *J. Am. Chem. Soc.*, 1933, **55**, 652.
24. M. J. Blandamer, J. Burgess, R. E. Robertson and J. M. W. Scott, *Chem. Rev.*, 1982, **82**, 259.
25. P. Molyneux, reference 13, Vol. 4, Chapter 7.
26. D. Eagland, reference 13, Vol. 4, Chapter 5.
27. M. L. Bender, *Mechanisms of Homogeneous Catalysis from Protons to Proteins*, Wiley-Interscience, New York, 1971.
28. C. A. Kraus, *J. Phys. Chem.*, 1956, **60**, 129.
29. G. Atkinson and Y. Mori, *J. Chem. Phys.*, 1966, **45**, 4716.
30. J. H. Sharp and M. C. R. Symons, *Ions and Ion-Pairs in Organic Reactions* (ed. by M. Szwarc), Wiley, 1972, Chapter 5.
31. S. G. Entelis and R. P. Tiger, *Reaction Kinetics in the Liquid Phase*, Wiley, New York, 1976.
32. C. K. Ingold, *Structure and Mechanism in Organic Chemistry*, G. Bell, London, 1953.
33. C. Reichardt, *Molecular Interactions* (ed. by H. Ratajczak and W. J. Orville-Thomas), Wiley, New York, 1982, Chapter 5.
34. E. Grunwald and S. Winstein, *J. Am. Chem. Soc.*, 1948, **70**, 846.
35. J. E. Leffler and E. Grunwald, *Rates and Equilibria of Organic Reactions*, Wiley, New York, 1963.
36. M. R. J. Dack, *Techniques of Chemistry* (ed. by A. Weissberger), Vol. VIII; *Solutions and Solubilities* (ed. by M. R. J. Dack), Part II, Wiley, New York, 1976, Chapter II.
37. J. B. Hyne, *Hydrogen-Bonded Solvent Systems* (ed. by A. K. Covington and P. Jones), Taylor & Francis, London, 1968, p. 69.

38. M. J. Blandamer, J. Burgess, P. P. Duce, R. E. Robertson and J. M. W. Scott, *J. Chem. Soc., Faraday Trans. I*, 1981, **77**, 1999.
39. M. J. Blandamer and J. Burgess, *Pure and Appl. Chem.*, 1983, **55**, 55.
40. C. A. Bunton and L. Robinson, *J. Am. Chem. Soc.*, 1968, **90**, 5965.
41. J. H. Fendler and E. J. Fendler, *Catalysis in Micellar and Macromolecular Systems*, Academic Press, New York, 1975.
42. M. J. Blandamer, J. Burgess and B. Clark, *J. Chem. Soc. Chem. Comm.*, 1983, 659.
43. I. D. Robb (Ed.), *Microemulsions*, Plenum Press, London, 1982.
44. M. J. Blandamer, J. Burgess, B. Clark, P. P. Duce and J. M. W. Scott, *J. Chem. Soc., Faraday Trans., II*, 1984, **80**, 739.

# 2

# SOLVENT EFFECTS IN ORGANIC CHEMISTRY

CHRISTIAN REICHARDT

*Department of Chemistry, Philipps University, Marburg, Federal Republic of Germany*

## ABSTRACT

*Studies of the effect of solvents on reaction rates (Berthelot and Saint-Gilles, Menshutkin) and on chemical equilibria (Claisen, Knorr and Wislicenus) are important milestones in the history of organic chemistry in the late 19th century. More recent developments comprise solvent effects on competitive reaction mechanisms, dichotomic reaction paths, chemoselectivity, stereoselectivity and absorption spectra, typical examples of which are given.*

*Solvent effects can be better understood if solvent polarity, a property which is not easy to define and to measure, is replaced by solvation capability of a solvent, empirical parameters of solvent polarity, e.g. the $E_T(30)$-scale. Examples of the application of this scale to solvent-sensitive chemical reactions are given.*

## 1. INTRODUCTION

In 1862, Marcelin Berthelot, Professor at the Collège de France in Paris, and his co-worker Léon Péan de Saint-Gilles carried out a remarkable experiment:[1]

They mixed together equivalent amounts of acetic acid and ethanol and measured the rate of esterification by determining, at distinct time intervals, the amount of the resulting ethyl acetate. In order to study the influence of temperature on the reaction rate, one part of the reaction mixture was put in a cool cellar, whereas the other part was kept in a warm laboratory room. The influence of the reaction volume was examined by diluting the reaction mixture with different inert organic solvents such as diethyl ether and benzene.

It happened that—at constant temperature, time and concentration—the reaction rate was very low in the ethereal solution, and only traces of ester could be detected. In benzene solution, however, a substantial amount of ester was formed.

In their classical paper 'Recherches sur les affinités', published in 1862,[1] Berthelot and Saint-Gilles stated that '... the esterificaton is disturbed and decelerated on the addition of neutral solvents not belonging to the reaction'. This was the first experimental observation showing that solvents may have an influence on chemical reactions.

But these interesting results seem to have been forgotten for many years. It was only in 1887 that Nikolai Menshutkin, Professor at the University of St Petersburg, Russia, again took up these investigations.

After thorough studies on the same esterification reaction and on the alkylation of tertiary amines with alkyl halides, he concluded, in 1890, that a reaction cannot be separated from the medium in which it is performed.[2] He found, for example, that a solvent change from $n$-hexane to benzyl alcohol for the $S_N2$ displacement reaction between triethylamine and ethyl iodide caused a 742-fold rate increase (cf. Fig. 1).

After strong efforts to find a correlation between this kind of solvent effect and distinct physical or chemical properties of the solvents used, he stated resignedly, in 1900, that there is only a very loose connection between the physical properties of organic solvents and their influence on reaction rates.

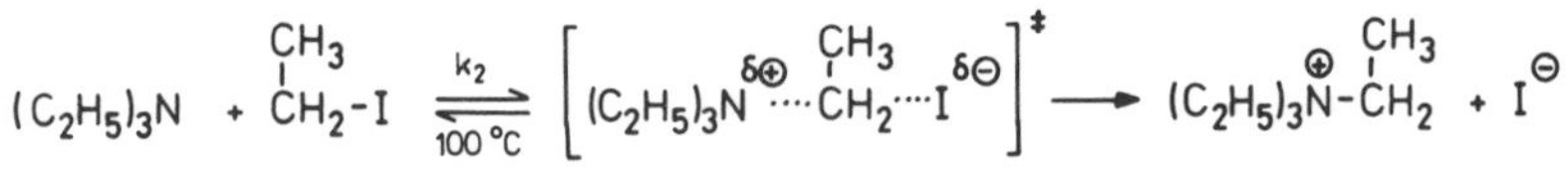

$$(C_2H_5)_3N \; + \; \overset{\overset{\displaystyle CH_3}{|}}{CH_2}\text{-}I \; \underset{100\,°C}{\overset{k_2}{\rightleftharpoons}} \; \left[(C_2H_5)_3\overset{\delta\oplus}{N}\cdots \overset{\overset{\displaystyle CH_3}{|}}{CH_2}\cdots \overset{\delta\ominus}{I}\right]^{\ddagger} \longrightarrow (C_2H_5)_3\overset{\oplus}{N}\text{-}\overset{\overset{\displaystyle CH_3}{|}}{CH_2} \; + \; I^{\ominus}$$

| Solvent | $n$-Hexane | Benzene | Methanol | Benzyl alcohol |
|---|---|---|---|---|
| $k_2^{rel}$ | 1 | 36 | 287 | 742 |
| solvent polarity | | | | ⟶ |

Fig. 1. Solvent effect on reaction rate: $S_N2$ displacement reaction of triethylamine with ethyl iodide.[2]

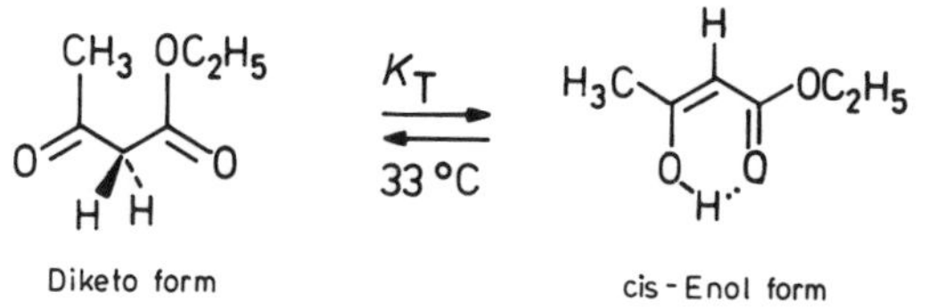

SOLVENT EFFECTS IN ORGANIC CHEMISTRY

| Solvent | Gas phase | n-Hexane | Benzene | Methanol | Acetic acid |
|---|---|---|---|---|---|
| $K_T$ | 0,74 | 0,64 | 0,19 | 0,062 | 0,019 |
| % Enol | 43 | 39 | 16 | 5,8 | 1,9 |
| solvent polarity | - - - - - - - - - | ———————————————→ | | | |

Fig. 2. Solvent effect on chemical equilibrium: keto–enol tautomerism of ethyl acetoacetate.[7,8]

This statement, even now, 120 years after the first discovery of solvent effects on reaction rates, is to some extent still valid.

The influence of solvents on the position of chemical equilibria was discovered for the first time in 1896 in Germany by Ludwig Claisen in Aachen,[3] Arthur Hantzsch in Würzburg,[4] Ludwig Knorr in Jena,[5] and Johannes Wislicenus in Würzburg,[6] simultaneously with the discovery of the keto–enol and the nitro–isonitro tautomerism.

A well-known example for the solvent effect on keto–enol tautomerism is shown in Fig. 2.[7,8]

The tautomeric equilibrium constants of ethyl acetoacetate indicate, for this cis-enolising β-dicarbonyl compound, an increasing enol content with decreasing solvent polarity.

## 2. OTHER TYPES OF SOLVENT EFFECTS ON CHEMICAL REACTIONS

The short historical introduction has already given us two examples of solvent effects on reaction rates and chemical equilibria. Dealing with solvent effects in organic chemistry includes, however, a greater variety of solvent effects. Only two further examples will be given.

First, a solvent can drastically change the mechanism of a reaction.

An example is the thermolysis of arene diazonium salts in solution where two competitive reaction mechanisms have been established (cf. Fig. 3).

In solvents of low nucleophilicity such as hexafluoroisopropanol (HFIP) or trifluoroethanol (TFE), the first and rate-determining step is a heterolytic cleavage of the diazonium ion to give an aryl cation and molecular nitrogen, followed by rapid reaction of the phenyl cation with any available nucleophile.

In solvents of high nucleophilicity and low redox potential such as pyridine or dimethylsulphoxide (DMSO), however, a homolytic decomposition of the diazonium ion is favoured to give an aryl radical and consequent reaction products.

Thus, depending on the solvent used, a substrate can react by two completely different reaction mechanisms, as shown by this de-diazoniation reaction. These investigations were carried out by Zollinger and his co-workers.[9]

A second example (shown in Fig. 4) demonstrates the solvent effect on chemoselectivity, that is on functional group differentiation.

The reduction of the bifunctional compound 11-bromoundecyl tosylate with lithium aluminium hydride in different solvents has been examined by Krishnamurthy.[10] In diethyl ether as solvent, the reaction proceeds with the selective reduction of the tosylate group to give $n$-undecyl bromide, whereas in diglyme as solvent, the bromide substituent is selectively reduced to yield $n$-undecyl tosylate. This

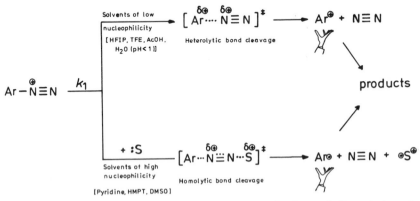

Fig. 3. Solvent effect on competitive reaction mechanisms: dediazoniation of arene diazonium salts.[9ab]

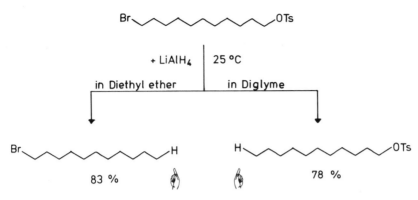

Fig. 4. Solvent effect on chemoselectivity: reduction of 11-bromoundecyl tosylate with LiAlH$_4$.[10]

reaction is an impressive example of how solvents can be used to modify the chemoselectivity of a complex metal hydride.

Thus, by a careful choice of the solvent, it is possible, using the same reagent, to selectively react the various functional groups in an organic molecule.

## 3. SOLVENT EFFECT ON REACTION RATES

In the light of the transition-state theory, solvent effects on rate constants depend on the relative stabilisation of the reactant molecules and the corresponding activated complex through solvation.

The early pioneering work of Hughes and Ingold from about 1933 onwards laid the foundation of a classification of organic reactions with respect to the expected solvent influence on the reaction rate.[11–13] Using a simple qualitative solvation model which allows only pure electrostatic interactions, they divided substitution and elimination reactions on the basis of different charge-type.

When the passage to the transition state involves the creation or concentration of ionic charges, the activated complex will be more strongly solvated than the reactants. An increase in solvent polarity will therefore favour the transition state relative to the initial state, and a rate acceleration results with increasing solvent polarity.

Conversely, when the activation process involves disappearance or dispersal of charges, the reactants will be solvated more strongly in the initial state than in the transition state. A rate deceleration with increasing solvent polarity will result in this case.

A change in solvent polarity will have a negligible effect on the rate of reactions which involve little or no change in the charge density in going from reactants to the activated complex.

A more general classification of organic reactions with respect to solvent effects, not restricted to substitution and elimination reactions, is shown in Fig. 5.[12,14,15]

According to this scheme, organic reactions can be roughly divided into three classes depending on the character of the activated complex involved: dipolar, isopolar, and free-radical transition-state reactions. Some typical examples to illustrate the usefulness of this classification will be given next. For practical purposes it is obviously important to know whether a solvent change will lead to a rate acceleration or not.

Dipolar transition-state reactions with large solvent effects can be found amongst ionisation, displacement, elimination and fragmentation reactions. Figures 6 and 7 show two typical examples.

A classical example for a dipolar transition-state reaction is the $S_N1$ solvolysis of *tert*-butyl chloride. The transition state of this reaction involves a partial separation of unlike charges (cf. Fig. 6).

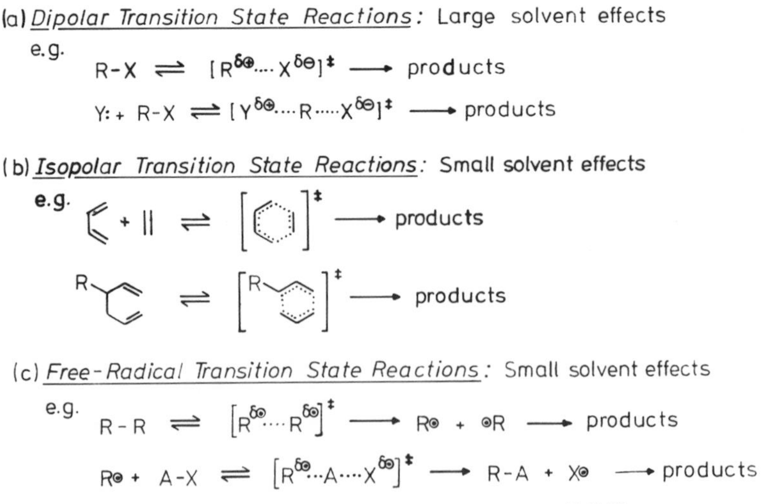

(a) *Dipolar Transition State Reactions:* Large solvent effects

e.g.

$$R\text{-}X \rightleftharpoons [R^{\delta\oplus}\cdots X^{\delta\ominus}]^{\ddagger} \longrightarrow products$$

$$Y\colon + R\text{-}X \rightleftharpoons [Y^{\delta\oplus}\cdots R\cdots X^{\delta\ominus}]^{\ddagger} \longrightarrow products$$

(b) *Isopolar Transition State Reactions:* Small solvent effects

e.g.

$\longrightarrow$ products

$\longrightarrow$ products

(c) *Free-Radical Transition State Reactions:* Small solvent effects

e.g.

$$R\text{-}R \rightleftharpoons [R^{\delta\ominus}\cdots R^{\delta\ominus}]^{\ddagger} \longrightarrow R\bullet + \bullet R \longrightarrow products$$

$$R\bullet + A\text{-}X \rightleftharpoons [R^{\delta\ominus}\cdots A\cdots X^{\delta\ominus}]^{\ddagger} \longrightarrow R\text{-}A + X\bullet \longrightarrow products$$

Fig. 5. Classification of organic reactions.[12,14,15]

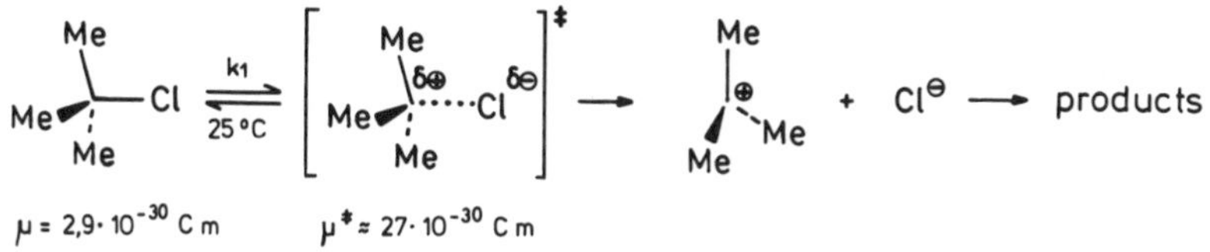

$\mu = 2,9 \cdot 10^{-30}$ C m          $\mu^{\ddagger} = 27 \cdot 10^{-30}$ C m

| Solvent | Benzene | Acetone | DMF | Methanol | Water |
|---|---|---|---|---|---|
| $k_1^{rel}$ | 1 | $1,9 \cdot 10^2$ | $4,8 \cdot 10^3$ | $1,1 \cdot 10^6$ | $4,2 \cdot 10^{10}$ |
| solvent polarity | | | | | |

Fig. 6. A dipolar transition-state reaction with large solvent effects: $S_N1$ solvolysis of *tert*-butyl chloride.[16,17]

The increased solvation of the dipolar activated complex relative to the less dipolar starting molecule leads to a huge rate acceleration with increasing solvent polarity. In going from benzene to water, the rate constants differ by a factor of $10^{10}$.[16,17] Protic solvents such as methanol and water play an additional role in this case: they act as a hydrogen-bond donating electrophile by solvating the incipient chloride ion.

In strong contrast to the ion-forming solvolysis of *tert*-butyl chloride, the $S_N2$ displacement reaction between chloride ions and uncharged methyl bromide is much slower in water than in dipolar aprotic solvents. This difference can be as large as $10^5$ (cf. Fig. 7).[18ab]

Due to the dispersal of negative charge during activation, the transition state is less solvated in solvents of increasing polarity. In going from dipolar aprotic to protic solvents, the chloride ion is further stabilised through hydrogen bonding. The result is that in water the initial reactants are much more stabilised than the large transition-state anion. Therefore, the free energy of activation is higher in water and the reaction is slower in this solvent.

This ionic $S_N2$ reaction is one of the rare examples which have been studied both in solution and in the gas phase.[19ab,20] Until recently, only solution-phase data were available for dipolar transition-state reactions. Solvent interference, however, makes it difficult to distinguish the intrinsic properties of the reactants from solvation effects! In the absence of any solvent molecules, the intrinsic reactivity of the naked

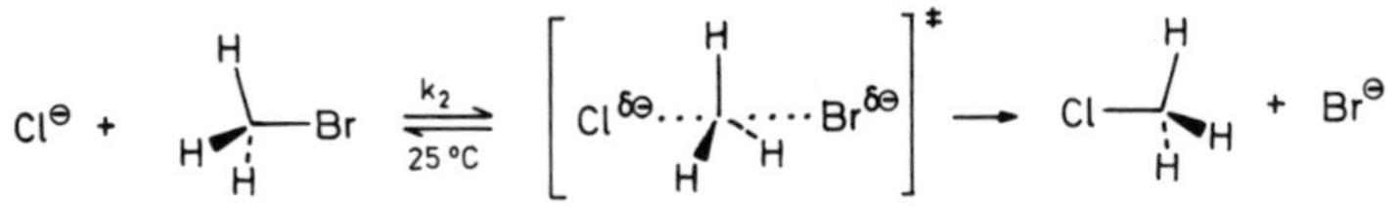

| Solvent | Protic solvents | | Dipolar aprotic solvents | | Gas phase |
|---------|-------|----------|-----|---------|-----------|
|         | Water | Methanol | DMF | Acetone | No solvent |
| $k_2^{rel}$ | 1 | 1,3 | $8 \cdot 10^4$ | $6 \cdot 10^5$ | $2 \cdot 10^{15}$ |
|         | ca. 1 | : | ca. $10^5$ | : | ca. $10^{15}$ |
| $E_a$ [kJ/mol] | 103 | | | | 11 |
| solvent polarity | | | | | |

Fig. 7. A dipolar transition-state reaction with large solvent effect: $S_N2$ displacement reaction of methyl bromide with the chloride ion.[18-20]

reactants can be measured and distinguished from the effects attributable to solvation.[19ab,20] The result in this case is truly remarkable. In the gas phase, the reaction of chloride ions with methyl bromide is faster than it is in water by a factor of $10^{15}$!

These and other results lead to the conclusion that reaction rates in solution are primarily determined by the energy needed to destroy the solvation shell during the activation process, and only to a minor extent by the intrinsic properties of the reactants. In other words, in solution reactions the solvent is almost solely responsible for the observed rate constants. The chemistry in solution is mainly determined by the solvents and not by the reactants.

Another class of organic reactions showing only small solvent effects is isopolar transition-state reactions (cf. Fig. 8). By definition, isopolar activated complexes differ very little in charge distribution from the initial reactants. Isopolar transition state reactions can be found in pericyclic reactions such as cycloaddition, sigmatropic, electrocyclic, and cheletropic reactions. One example illustrates the lack of solvent sensitivity of pericyclic reactions (cf. Fig. 8).

In the Diels–Alder [4 + 2] cycloaddition reaction of 2,3-dimethylbutadiene with 4-chloro-1-nitrosobenzene, only a very small solvent effect, a factor of two, has been found.[21] This supports the formulated one-step mechanism with synchronous bond formation without any

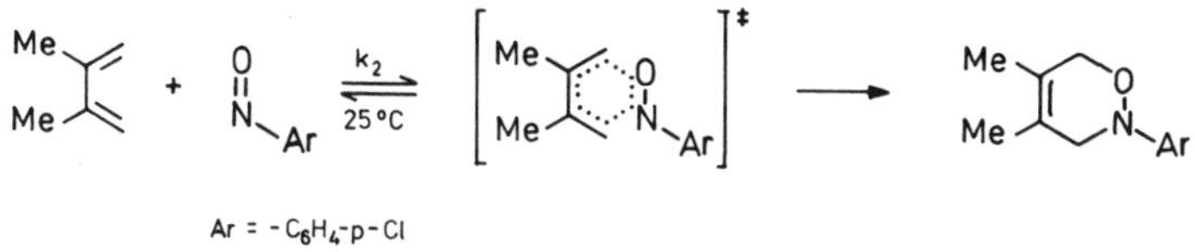

Ar = -C$_6$H$_4$-p-Cl

| Solvent | Toluene | Dichloromethane | Ethanol | Nitrobenzene |
|---|---|---|---|---|
| $k_2^{rel}$ | 1,0 | 1,4 | 2,0 | 2,2 |
| $\Delta G^{\ddagger}$[kJ/mol] | 87,0 | 86,2 | 85,4 | |
| solvent polarity | | | | ⟶ |

Fig. 8. An isopolar transition-state reaction with small solvent effect: [4 + 2] cycloaddition reaction of 2,3-dimethylbutadiene with 4-chloro-1-nitroso-benzene.[21]

creation, destruction, or dispersal of charge during the activation process.

A third category of organic reactions involves free-radical transition states, formed by the creation of unpaired electrons during radical-pair formation or atom-transfer reactions (cf. Fig. 9). In reactions of this type only negligible solvent effects have normally been observed. No creation, destruction, or dispersal of charge is connected with the activation process, which involves the homolytic cleavage of one or more covalent bonds.

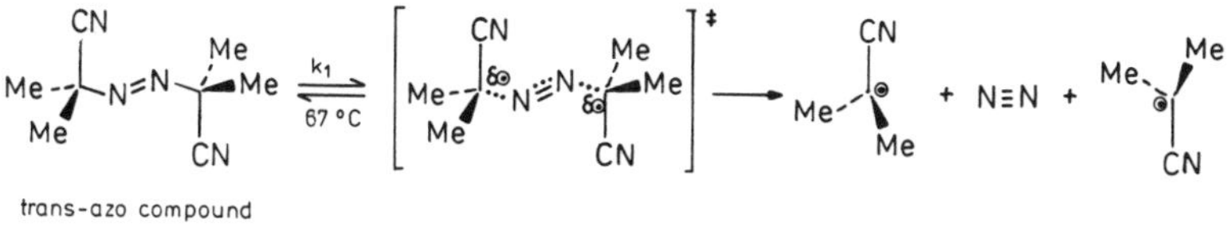

trans-azo compound

| Solvent | Decalin | Chlorobenzene | o-Dichlorobenzene | N,N-Dimethylaniline |
|---|---|---|---|---|
| $k_1^{rel}$ | 1,0 | 1,4 | 1,5 | 1,6 |
| $E_a$[kJ/mol] | 148 | 133 | 131 | 128 |
| solvent polarity | | | | ⟶ |

Fig. 9. A free-radical transition-state reaction with small solvent effect: thermolysis of azobisisobutyronitrile.[22]

One typical example of a free-radical transition-state reaction is shown in Fig. 9.

The thermolysis of azobisisobutyronitrile, an apolar *trans*-azo compound, is not sensitive to medium effects as shown by the nearly equal relative first-order rate constants measured in different solvents.[22] This is in agreement with the homolytic, concerted two-bond scission mechanism, producing two neutral radicals as intermediates without any charge separation.

In conclusion, it can be stated that a great variety of organic reactions can be reasonably classified with respect to their solvent sensitivity. Vice versa, the solvent influence on a reaction under study can be an important criterion for the elucidation of its mechanism.

## 4. EMPIRICAL PARAMETERS OF SOLVENT POLARITY

Having discussed the solvent sensitivity of certain organic reactions, the question remains, which solvent property is responsible for the medium effects, and whether it is possible to describe these solvent effects in a more quantitative manner.

Organic chemists usually attempt to understand solvent effects in terms of the 'polarity of the solvent'. It is, however, not easy to define this property precisely, and it is even more difficult to assess it quantitatively.

Seduced by the simplicity of electrostatic solvation models, attempts at expressing solvent polarity quantitatively usually involve physical solvent properties such as dielectric constant, dipole moment, or refractive index. But this procedure is often inadequate, particularly because it does not take into account specific solute–solvent interactions such as hydrogen bonding, charge-transfer forces, solvophobic interactions, etc.

Hence, from a more practical point of view, it seems reasonable to understand solvent polarity in terms of the overall solvation capability. This solvation capability of a solvent is in turn determined by the sum of *all* those molecular properties responsible for the solute–solvent interactions.[12,13] It is obvious that solvent polarity, thus defined, cannot be described quantitatively by a single physical parameter.

In such a situation other indices of solvent polarity are sought. The lack of comprehensive theoretical expressions for the calculation of solvent effects, and the inadequacy of defining solvent polarity in terms of simple physical constants, has led to the introduction of purely empirical scales of solvent polarity. Based on the assumption that particular solvent-dependent chemical reactions or spectral absorptions may serve as suitable model processes, reflecting all the possible solute–solvent interactions in the solvation game, several empirical parameters of solvent polarity have been introduced recently by different authors.[23–26]

This approach is very characteristic of the experimental chemist, but it is regarded with mistrust by theoretical chemists. However, organic chemistry has always made use of the qualitative, empirical rule that similar compounds react in similar ways, and that similar changes in the structure or in the reaction medium produce similar changes in the chemical reactivity.

One of the first verifications of this rule was the introduction of the so-called Hammett equation for the calculation of substituent effects on chemical reactivity, using the ionisation reaction of substituted benzoic acids in water at 25 °C as a reference process.[27]

The various model processes used to establish such empirical scales of solvent polarity have been reviewed.[12,23–26]

Only one of these parameters, introduced by us some years ago as the so-called $E_T(30)$-scale,[28abc,29] will be mentioned. By virtue of its exceptionally large negative solvatochromism, the pyridinium-N-phenoxide betaine dye shown in Fig. 10 is particularly suitable as a standard dye for the determination of an empirical solvent parameter.[30]

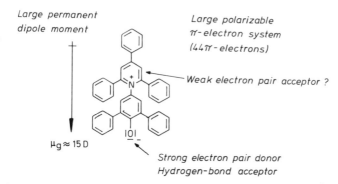

Large permanent dipole moment

$\mu_g \approx 15\,D$

Large polarizable π-electron system (44π-electrons)

Weak electron pair acceptor ?

Strong electron pair donor
Hydrogen-bond acceptor

Fig. 10. Ground state properties of the negatively solvatochromic pyridinium-N-phenoxide betaine dye used to establish the $E_T(30)$-scale (dye no. 30 in Ref. 28abc).

This betaine dye exhibits:

(i)  a large permanent dipole moment of about 15 Debyes, suitable for registration of dipole/dipole and dipole/induced-dipole interactions between solute and solvent;

(ii)  a large polarisable π-electron system, consisting of 44 π-electrons, suitable for the registration of dispersion interactions;

(iii)  the phenolic oxygen atom represents a highly basic electron-pair donor centre, suitable for specific hydrogen-bond interactions with protic solvents.

With increasing solvent polarity, the long-wavelength absorption band of this betaine dye is shifted hypsochromically due to the increasing stabilisation of the dipolar electronic ground state relative to the less dipolar first excited state. The long-wavelength solvatochromic absorption band is situated at 810 nm in diphenyl ether and 453 nm in water as solvents (cf. Fig. 11).[28abc]

With the corresponding solvent-induced hypsochromic shift of 357 nm, this betaine dye still holds the world record in solvatochromism.

The molar transition energies, $E_T$, of this betaine dye measured in kcal/mol, have been proposed by us as an empirical parameter of

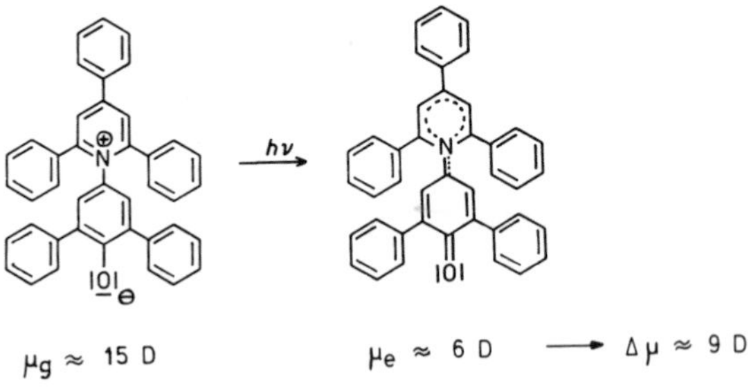

$\mu_g \approx 15\ D$          $\mu_e \approx 6\ D$  ⟶  $\Delta\mu \approx 9\ D$

| Solvent | $C_6H_5OCH_3$ | $CH_3COCH_3$ | $i\text{-}C_5H_{11}OH$ | $C_2H_5OH$ | $CH_3OH$ |
|---|---|---|---|---|---|
| λ [nm] | 769 | 677 | 608 | 550 | 515 |
| $E_T$ [kcal/mol] | 37.2 | 42.2 | 47.0 | 51.9 | 55.5 |
| Solution colour | yellow | green | blue | violet | red |

$$E_T\,[\text{kcal/mol}] \equiv h\cdot c\cdot \tilde{\nu}\cdot N = 2.859\cdot 10^{-3}\cdot \tilde{\nu}\ [\text{cm}^{-1}]$$

Fig. 11. Solvent-dependent intramolecular charge-transfer absorption of the standard pyridinium-N-phenoxide betaine dye.

solvent polarity.[28abc] A high $E_T$-value corresponds to high solvent polarity. At present, $E_T(30)$-values for nearly 250 organic solvents and for many binary solvent mixtures are available.[29]

Since the greater part of the solvatochromic absorption range lies within the visible region, it is even possible to make a visual estimate of the solvent polarity.

The solution colour of this betaine dye is red in methanol, violet in ethanol, green in acetone, blue in isoamyl alcohol, and greenish-yellow in anisole. With suitable binary solvent mixtures, almost every colour of the visible spectrum can be obtained.

There are some limitations of the $E_T$-scale:

(i)   no $E_T$-values can be obtained for acidic solvents. Protonation of the phenolic oxygen atom of the betaine dye leads to disappearance of the solvatochromic absorption band;

(ii)  due to the low volatility of the betaine dye, no gas-phase $E_T$-value can be directly determined;

(iii) $E_T$-values for non-polar solvents such as hydrocarbons cannot be directly measured, due to the low solubility of the standard betaine dye in those solvents.

The last problem can be overcome by the use of alkyl-substituted betaine dyes such as shown in Fig. 12.[29]

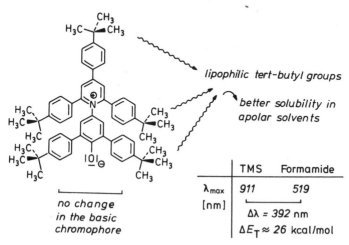

| | TMS | Formamide |
|---|---|---|
| $\lambda_{max}$ [nm] | 911 | 519 |

$\Delta\lambda = 392$ nm

$\Delta E_T \approx 26$ kcal/mol

lipophilic tert-butyl groups

better solubility in apolar solvents

no change in the basic chromophore

Fig. 12. A new secondary penta-tert-butyl-substituted pyridinium-N-phenoxide betaine dye for the determination of $E_T(30)$-values for nonpolar solvents.[29]

The introduction of five *tert*-butyl groups into the peripheral phenyl groups does not change the basic chromophore, but does yield a more lipophilic betaine dye.

The excellent linear correlation between the $E_T$-values of both the standard betaine dye and its penta-*tert*-butyl-substituted derivative allows the calculation of $E_T$-values for solvents in which our standard betaine dye is not soluble.[29]

Unfortunately, $E_T$-values have by definition the dimensions of kcal/mol,[28abc] a unit which should be abandoned in the framework of SI units.

We have recently recommended, therefore, the use of the so-called normalised $E_T^N$-values.[29] These are defined according to eqn (2) in Fig. 13 as the quotient of two differences: the difference between the $E_T$-value of the solvent under consideration and of tetramethylsilane, and the difference between the $E_T$-value of water and of tetramethylsilane.

Hence, the corresponding $E_T^N$-scale ranges from 0·0 for tetramethylsilane, the least polar solvent, to 1·0 for water, the most polar solvent (cf. Fig. 13). These $E_T^N$-values are thus dimensionless numbers such as Hammett's substituent constants.[27] They are known for nearly 250 organic solvents.[29] $E_T$-values have found wide application in correlating and predicting solvent effects on reaction rates, on chemical equilibria, on the position of absorption bands in UV/Vis-, IR-, ESR- and NMR-spectroscopy.[12,23–25,31]

$$E_T \ [kcal/mol] \equiv h \cdot c \cdot \tilde{\nu} \cdot N = 2{,}859 \cdot 10^{-3} \cdot \tilde{\nu} \ [cm^{-1}] \qquad (1)$$

$$E_T^N \equiv \frac{E_T(\text{Solvent}) - E_T(\text{TMS})}{E_T(\text{Water}) - E_T(\text{TMS})} = \frac{E_T(\text{Solvent}) - 30{,}7}{32{,}4} \qquad (2)$$

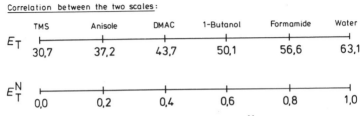

Fig. 13. Definition of $E_T$-values and normalised $E_T^N$-values according to eqns (1) and (2), respectively.[29]

## 5. APPLICATIONS OF THE $E_T(30)$-SCALE

Since solvent polarity parameters such as the $E_T$-scale are established in an empirical way, their usefulness has to be tested in the same manner.

Therefore, finally, two examples for the application of our $E_T$-parameter will be given.

The $S_N1$ solvolysis of *tert*-butylchloride has been already mentioned (cf. Fig. 6). Figure 14 shows the correlation between our $E_T$-values and the ionisation rate of *tert*-butyl chloride, measured in 17 solvents.[16,17]

The linear correlation between the solvent polarity parameter and the rate of this typical dipolar transition-state reaction is so good that rate constants for further solvents can easily be calculated.

This astonishingly good correlation between the spectroscopically determined solvent polarity parameter and the kinetically determined

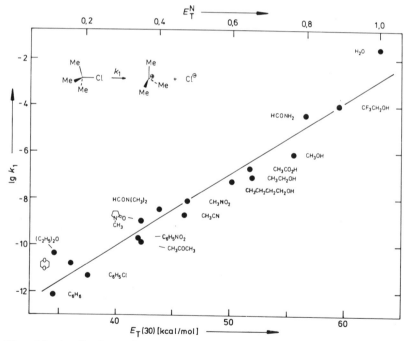

Fig. 14. Application of solvent polarity parameters: Linear correlation between $E_T(30)$-values and $\log_{10} k_1$ of the $S_N1$ solvolysis of *tert*-butyl chloride at 25 °C.[16,17] ($\log_{10} k_1 = 0\cdot312E_T - 22\cdot55$; $n = 17$; $r = 0\cdot972$)

rate constant, shows that the light absorption of our solvatochromic dye is indeed a good model process for the estimation of solute–solvent interactions in dipolar transition-state reactions.

In spite of the small solvent effects, even isodipolar transition-state reactions such as the Diels–Alder cycloaddition reaction of 2,3-dimethylbutadiene with 4-chloro-1-nitrosobenzene,[21] mentioned before (cf. Fig. 8), exhibit a satisfactory linear correlation with the solvent polarity parameter $E_T$ (cf. Fig. 15).

The gentle slope of this correlation line is typical for concerted cycloaddition reactions with a one-step mechanism involving little or no dipolarity change during the activation process. Similar gentle slopes have been observed for other pericyclic reactions such as the 1,3-dipolar cycloaddition reactions mainly studied by Huisgen and co-workers.[32]

For an estimate of solvent effects on chemical reactions, empirical parameters of solvent polarity are as useful as the Hammett constants

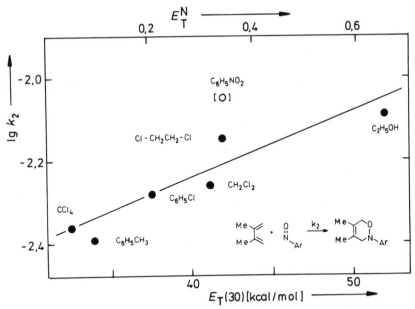

Fig. 15. Application of solvent polarity parameters: linear correlation between $E_T(30)$-values and $\log k_2$ of the $[4 + 2]$ cycloaddition reaction of 2,3-dimethyl-butadiene with 4-chloro-1-nitrosobenzene[21] ($\log k_2 = 0.016 E_T - 2.88$; $n = 6$; $r = 0.936$).

for the prediction of substituents' effects on chemical reactions. One important reservation should be made. Solvent effects are basically more complicated and more specific than substituents' effects.

The application of solvent polarity parameters is based on the assumption that the nature of the solute–solvent interactions in the reference process, used to establish a solvent scale, is similar to those influencing the reaction under study. This is obviously true only for closely related solvent-sensitive reactions. Therefore, the use of solvent parameters to predict solvent effects should be limited to analogous processes.

A large variety of empirical parameters of solvent polarity are presently at the chemist's disposal.[12,23-26] Therefore, it should not be too difficult to find a suitable solvent parameter for a particular reaction under study.

During the last few years there have been various attempts at considering two or more specific solute–solvent interactions simultaneously, using multiparameter equations instead of single-parameter correlations. Different empirical parameters for solvent polarisability, dipolarity, electrophilicity, and nucleophilicity have been separately determined and then combined in multiparameter equations (cf. Fig. 16).[25,26,33-35]

The use of multiparameter equations instead of single-parameter equations has in many cases produced a dramatic improvement in the correlations between solvent-dependent reactions and inherent solvent properties.[26,33-35] However, a final solution to the problem of

$$A = A_0 + b \cdot B + c \cdot C + d \cdot D + \ldots$$

$A$ = Solvent-dependent physicochemical quantity ($\lg K$, $\lg k$, $E_T$, etc.), determined in a series of given solvents.

$A_0$ = Corresponding property in the gas phase or in an inert reference solvent.

$B, C, D, \ldots$ = Independent but complementary solvent parameters which account for the different solute–solvent interaction mechanisms.

$b, c, d, \ldots$ = Regression coefficients describing the sensitivity of property $A$ to the respective solute–solvent interaction.

Fig. 16. Multiparameter approach to solvent effects.

separating solvent polarity into the various components of solute–solvent interactions is not yet in sight.

## 6. CONCLUSION

One hundred and twenty years after the discovery by Berthelot and Saint-Gilles of the solvent influence on chemical reactivity, the problem of how to understand and describe the solvent influence on chemical reactivity in a more quantitative manner is far from being finally solved.

Most elementary organic reactions can be classified into dipolar, isopolar, and free-radical transition-state reactions according to their solvent sensitivity. On this basis, the extent and direction of a solvent influence on reaction rates can be used as a valuable criterion in the elucidation of reaction mechanism. Thereby, extreme rather than intermediate solvent effects are more conclusive.

The use of empirical parameters of solvent polarity is still the simplest and most practical method of predicting solvent effects in a more quantitative way. Various empirical solvent scales are now available for this procedure. Multiparameter equations have led to further improvements.

As the selection of suitable reference processes for the determination of empirical parameters of solvent polarity leaves room for subjective decisions, the well-known remark of Hammett[27] that organic chemistry is not a science but an art, seems to be not wholly unfounded.

## REFERENCES

1. M. Berthelot and L. Péan de Saint-Gilles, *Ann. Chim. et Phys., 3. Sér.,* 1862, **65,** 385–422; *ibid.,* 1862, **66,** 5–110; *ibid.,* 1863, **68,** 225–359; translated into German and ed. by M. and A. Ladenburg in: *Ostwald's Klassiker der exakten Naturwissenschaften,* No. 173, Engelmann, Leipzig, 1910.
2. N. Menshutkin, *Z. Phys. Chem.,* 1887, **1,** 611–30; *ibid.,* 1890, **5,** 589–600; *ibid.,* 1890, **6,** 41–57; *ibid.,* 1900, **34,** 157–67.
3. L. Claisen, *Liebigs Ann. Chem.,* 1896, **291,** 25–111.
4. A. Hantzsch and O. W. Schultze, *Ber. Deutsch. Chem. Ges.,* 1896, **29,** 2251–67.
5. L. Knorr, *Liebigs Ann. Chem.,* 1896, **293,** 70–120.

6. W. Wislicenus, *Liebigs Ann. Chem.*, 1896, **291**, 147–216.
7. K. H. Meyer, *Liebigs Ann. Chem.*, 1911, **380**, 212–42; *Ber. Deutsch. Chem. Ges.*, 1914, **47**, 826–32.
8. M. T. Rogers and J. L. Burdett, *J. Am. Chem. Soc.*, 1964, **86**, 2105–9; *Can. J. Chem.*, 1965, **43**, 1516–26.
9. (a) I. Szele and H. Zollinger, *Helv. Chim. Acta*, 1978, **61**, 1721–9; (b) H. Zollinger, *Angew. Chem.*, 1978, **90**, 151–60; *Angew. Chem., Int. Ed. Engl.*, 1978, **17**, 141.
10. S. Krishnamurthy, *J. Org. Chem.*, 1980, **45**, 2550–1.
11. C. K. Ingold, *Structure and Mechanism in Organic Chemistry* (2nd edn), Cornell University Press, Ithaca, London, 1969.
12. C. Reichardt, *Solvent Effects in Organic Chemistry*, Verlag Chemie, Weinheim, New York, 1979, p. 85.
13. J. Shorter, *Correlation Analysis of Organic Reactivity*, Research Studies Press, Chichester, 1982, p. 127.
14. E. M. Kosower, *An Introduction to Physical Organic Chemistry*, Wiley, New York, 1968.
15. C. Reichardt, *Pure Appl. Chem.*, 1982, **54**, 1867–84.
16. S. Winstein and A. H. Fainberg, *J. Am. Chem. Soc.*, 1956, **78**, 2770–7; *ibid.*, 1957, **79**, 5937–50.
17. M. H. Abraham, *J. Chem. Soc., Perkin Trans. II*, 1972, 1343–57.
18. (a) R. Alexander, E. C. F. Ko, A. J. Parker and T. J. Broxton, *J. Am. Chem. Soc.*, 1968, **90**, 5049–69; (b) A. J. Parker, *Chem. Rev.*, 1969, **69**, 1–32.
19. (a) D. K. Bohme, G. I. Mackay and J. D. Payzant, *J. Am. Chem. Soc.*, 1974, **96**, 4027–8; (b) K. Tanaka, G. I. Mackay, J. D. Payzant and D. K. Bohme, *Can. J. Chem.*, 1976, **54**, 1643–59.
20. J. I. Brauman, W. N. Olmstead and C. A. Lieder, *J. Am. Chem. Soc.*, 1974, **96**, 4030–1.
21. G. Swieton and H. Kelm, *J. Chem. Soc., Perkin Trans. II*, 1979, 519–24.
22. M. G. Kulkarni, R. A. Mashelkar and L. K. Doraiswamy, *Chem. Eng. Sci.*, 1980, **35**, 823–30.
23. C. Reichardt, *Angew. Chem.*, 1979, **91**, 119–31; *Angew. Chem., Int. Ed. Engl.*, 1979, **18**, 98–110.
24. C. Reichardt, Solvent scales and chemical reactivity, in: *Organic Liquids. Structure, Dynamics, and Chemical Properties*, ed. by A. D. Buckingham, E. Lippert, and S. Bratos, Wiley, Chichester, New York, 1978, pp. 269–91.
25. C. Reichardt, Empirical parameters of solvent polarity and chemical reactivity, in: *Molecular Interactions*, ed. by H. Ratajczak and W. J. Orville-Thomas, Wiley, Chichester, New York, 1982, Vol. 3, pp. 241–82.
26. J. L. M. Abboud, M. J. Kamlet and R. W. Taft, *Progr. Phys. Org. Chem.*, 1981, **13**, 485–630.
27. L. P. Hammett, *J. Am. Chem. Soc.*, 1937, **59**, 96–103; *Physical Organic Chemistry*, McGraw-Hill, New York, 1970.
28. (a) K. Dimroth, C. Reichardt, T. Siepmann and F. Bohlmann, *Liebigs Ann. Chem.*, 1963, **661**, 1–37; (b) K. Dimroth and C. Reichardt, *ibid.*, 1969, **727**, 93–105; (c) C. Reichardt, *ibid.*, 1971, **752**, 64–7.

29. C. Reichardt and E. Harbusch-Görnert, *Liebigs Ann. Chem.*, 1983, 721–43.
30. C. Reichardt, E. Harbusch and R. Müller, Pyridinium-*N*-phenoxide betaine dyes as solvent polarity indicators. Some new findings, in: *Advances in Solution Chemistry*, ed. by I. Bertini, L. Lunazzi and A. Dei, Plenum Press, New York, London, 1981, pp. 275–93.
31. C. Reichardt and K. Dimroth, *Fortschr. Chem. Forsch.*, 1968, **11**, 1–73.
32. R. Huisgen, *Pure Appl. Chem.*, 1980, **52**, 2283–2302.
33. I. A. Koppel and V. A. Palm, The influence of the solvent on chemical reactivity, in: *Advances in Linear Free Energy Relationships*, ed. by N. B. Chapman and J. Shorter, Plenum Press, London, New York, 1972, pp. 203–80.
34. T. M. Krygowski and W. R. Fawcett, *J. Am. Chem. Soc.*, 1975, **97**, 2143; *Aust. J. Chem.*, 1975, **28**, 2115; *Can. J. Chem.*, 1976, **54**, 3283.
35. C. G. Swain, M. S. Swain, A. L. Powell and S. Alunni, *J. Am. Chem. Soc.*, 1983, **105**, 502.

# 3

# ROLE OF SOLVENTS IN INDUSTRY*

CHARLES W. ANDREWS

*Solvents Consultant, Dorking, Surrey, UK*

## ABSTRACT

*In broad terms, solvents are used in industry:*

1. *to remove a solid or liquid component from a mixture to be followed by purification;*
2. *to remove contaminants from metals, textiles and other substrates;*
3. *as carriers for solid components designed to protect and decorate fixed and mobile substrates.*

*The requirements of the operations above are many and involve considerations of water miscibility, solvent power, rates of evaporation, handling and health hazards, solvent recovery and the disposal of waste arising from such operations.*

*To satisfy all of these requirements, manufacturers of solvents have made available world-wide in bulk quantities more than 50 individual solvents along with a number of solvent mixtures, many of which arise as by-products from major processes.*

*This paper deals mainly with the major individual solvents, and sectors of industry using these solvents are identified and the reasons given for their choice. Some comments are included on the interchangeability of solvents and their standards of purity, their value to industry and their future role.*

---

* Presented at the 3rd Solvents Symposium in 1980 and revised in 1983.

41

# 1. INTRODUCTION

Until the introduction of petrochemistry, solvents such as acetone and
n-butanol were derived by fermentation processes. Petrochemical
solvents are now so well established that they are vital to most sections
of industry. More than 50 such solvents are available world-wide in
bulk quantities of qualities which are able to meet the most exacting
requirements.

Solvents used by industry are simple membered, stable and
non-corrosive chemical compounds usually of very low viscosity. Costly
large-scale plants produce solvents of exceptional purity, in some cases
down to 20/30 parts per million of total impurities. Thus international
standards for common solvents are usually set very high.

Table 1 lists the most common solvents used in industry, together
with some of their simple properties such as miscibility with water,
self-odour ratings, levels of solvent vapour just detectable in air,
threshold limit values, evaporation rates and flash points. A large
number of solvent mixtures are also marketed. They usually result
from by-products of chemical processes and are not featured in this
chapter.

This chapter discusses the various factors involved in choosing a
solvent and outlines the main solvents that are used in a number of
industrial operations.

## TABLE 1
### Common Industrial Solvents

| Solvent | Water misc.[a] | Odour[b] | TLV,[c] ppm | DLS,[d] ppm | Evap. rate | Boiling point, °C | Flash point,[e] °C |
|---|---|---|---|---|---|---|---|
| Ketones |  |  |  |  |  |  |  |
| Acetone | M | S | 1000 | 1000 | 1000 | 56 | −17 |
| Methyl ethyl | P | S | 200 | 150 | 480 | 80 | −7 |
| Methyl isobutyl | I | S | 100 | 25 | 160 | 116 | 13 |
| Isophorone | I | S | 5 | 30 | 3 | 215 | 96 |
| Diacetone alcohol | M | M | 50 | 25 | 9 | 168 | 60 |
| Cyclohexanone | P | S | 50 | 12 | 30 | 157 | 41 |
| Acetates |  |  |  |  |  |  |  |
| Ethyl | P | M | 400 | 100 | 500 | 77 | −5 |
| n-Butyl | I | M | 150 | 50 | 100 | 127 | 23 |
| Ethyl glycol ether | P | M | 100 | 25 | 20 | 156 | 58 |

TABLE 1 (*continued*)

| Solvent | Water misc.[a] | Odour[b] | TLV,[c] ppm | DLS,[d] ppm | Evap. rate | Boiling point, °C | Flash point, °C |
|---|---|---|---|---|---|---|---|
| **Alcohols** | | | | | | | |
| Methyl | M | W | 200 | 2000 | 400 | 65 | 12 |
| Ethyl | M | W | 1000 | 500 | 250 | 78 | 12 |
| Isopropyl | M | M | 400 | 400 | 200 | 82 | 12 |
| n-Butyl | P | S | 50 | 25 | 35 | 118 | 35 |
| **Glycol** | | | | | | | |
| Ethylene | M | W | 100 | — | <1 | 195 | 96 |
| **Glycol ethers** | | | | | | | |
| Ethyl | M | M | 100 | 200 | 25 | 136 | 49 |
| Butyl | M | M | 50 | 12 | 5 | 171 | 74 |
| **Hydrocarbons** | | | | | | | |
| 1. Aliphatics—aromatics up to 2% | | | | | | | |
| Hexane 62/68 | I | W | 100 | — | 850 | 62/68 | <−18 |
| SBP2 70/90 | I | W | 250 | — | 550 | 70/90 | <−18 |
| SBP5 90/100 | I | W | 300 | — | 320 | 90/100 | <−18 |
| SBP3 100/120 | I | W | 300 | — | 210 | 100/120 | −7 |
| Kerosene 195/260 | I | W | — | — | <2 | 195/260 | 70 |
| 2. Aliphatics—aromatics 2–25% | | | | | | | |
| SBP6 130/160 | I | M | 200 | — | 40 | 130/160 | 28 |
| White spirit 150/195 | I | M | 100 | — | 13 | 150/195 | 39 |
| Distillate 165/230 | I | W | — | — | 4 | 165/230 | 49 |
| 3. Aromatics | | | | | | | |
| Benzene | I | S | 10 | 75 | 500 | 80 | −11 |
| Toluene | I | S | 100 | 100 | 210 | 111 | 4 |
| Xylene | I | S | 100 | 50 | 70 | 139/144 | 29 |
| High boiling mixes | I | S | 25 | — | <50 | — | — |
| 4. Halogenated | | | | | | | |
| Methylene chloride | I | M | 200 | — | — | 40 | NF |
| Trichlorethylene | I | M | 100 | — | — | 87 | NF |
| Perchlorethylene | I | M | 100 | — | — | 121 | NF |
| 1,1,1-Trichlorethane | I | M | 350 | — | — | 74 | NF |
| **Miscellaneous** | | | | | | | |
| Dimethylformamide | M | M | 10 | — | 17 | 153 | 67 |
| Furfural | P | S | 5 | — | — | 162 | 67 |
| Sulpholane | M | — | — | — | solid | 285 | 106 |
| Tetrahydrofuran | M | M | 200 | — | 700 | 66 | −14 |

[a] Water miscibility: M, miscible; P, partially; I, immiscible.
[b] Odour: S, strong; M, medium; W, weak.
[c] Threshold limit value, parts per million.
[d] Detectable level of solvent in air, parts per million.
[e] NF, Non-flammable.

## 2. CHOOSING A SOLVENT

Before discussing the role of solvents in industry, a short review is given of some of the properties by which solvents are judged.

### 2.1. General

While the technical merit of a solvent is an obvious requirement, knowledge of its availability and cost of storing is essential. The cost-effective index of a solvent can be influenced by the considerable differences of cost that exist between bulk and drum deliveries.

Another simple but major factor is the operator and customer acceptance of a solvent, and here self-odour, flammability and health risks are often overriding considerations.

### 2.2. Solvent Power

This is a measure of the ability of a solvent to dissolve the solute. Clearly it is a most important property.

As a class, ketones are the most powerful of solvents, being approximately 10% stronger than the corresponding alkyl acetates. Halogenated and aromatic hydrocarbons also have strong solvent properties while aliphatic hydrocarbons are weaker. Alcohols are relatively weak solvents and feature more in chemical processes.

In a number of sectors of industry, a single solvent is adequate for the purpose, and choice is a relatively simple matter. Manufacturers of surface coatings, however, have problems which arise because of the substrate and the mode of application to that substrate. Whereas drying oil resin decorative coatings are usually based on white spirit, coatings for industrial production lines have to meet exacting drying and application standards which often require a balanced blend of solvents.

### 2.3. Evaporation Rate

The speed at which a solvent evaporates to the atmosphere is known as its evaporation rate and it is broadly in line with its distillation range.

Air humidity has little effect on the evaporation rates of non-polar solvents but at ambient temperatures, high boiling-point water-miscible solvents absorb atmospheric moisture during evaporation, causing their rates to slow. Air velocity across the surface of a solvent speeds evaporation significantly as, of course, does heat.

n-Butyl acetate is the widely accepted standard for evaporation rate against which other solvents are judged. It is given a reference value of 100 and, as can be seen from Table 1, at 25 °C, acetone has an approximate value of 1000 which means that it evaporates 10 times faster than n-butyl acetate and in turn approximately 70 times faster than white spirit.

It should be noted that the evaporation rates quoted are for a single solvent evaporating under strictly controlled conditions. Solutes can affect these time-related values in different ways, while in mixtures of solvents, azeotropes can form during evaporation and these have their own values.

## 2.4. Solvent Retention

True solvents are retained more readily in solutes than partial or non-solvents. Incomplete evaporation can cause problems in wrapping paper and film which are coated and printed. Surface dry coatings or print can still contain 10% of the original content of solvent; thus, stacking of coated paper or film which is not completely dry traps retained solvent, leading to residual odours. This may not be a problem if a food-grade solvent is used.

In bond-to-bond adhesives, retained solvent prevents full bond strength being achieved.

## 2.5. Flow and Levelling Properties

These factors feature in industrial coatings and printing inks. To obtain maximum durability and to optimise opacity, it is essential that newly applied films, which are usually very uneven, are allowed to flow out completely before setting. To achieve this in industrial coatings, a balanced blend of solvents is often required which will also take into account the many contours and angles of the article to be coated.

## 2.6. Safety Controls

As the use of solvents widened and increased, it became necessary to impose Statutory Regulations to cover aspects of manufacture, transportation, storage, handling and disposal. All such aspects must be taken into account when choosing a solvent.

Since solvents used by industry are relatively pure simple chemicals, there has been little difficulty in achieving international agreement on safety standards. The American Conference of Governmental Industrial Hygienists has been to the forefront in collecting relevant information on chemical substances in common use. Annually, they

publish threshold limit values of airborne concentrations of substances
which represent conditions under which it is believed that humans can
be exposed to daily without adverse effect.

Fire hazards are well known to most in industry and, within the
industry, the accepted classes of HIGHLY INFLAMMABLE, IN-
FLAMMABLE and NON-FLAMMABLE also have a bearing on
solvent choice.

## 3. ROLE OF SOLVENTS

That some 50 or so solvents are available in bulk quantities, is a
measure of the numerous and sometimes highly complex operations
that industry undertakes.

If the operation is one of extraction, e.g. isolating a pharmaceutical
or chemical intermediate, removing vegetable oils from crushed seeds
or cleansing metal sheet or parts, then often a single solvent is
adequate and recovery of that solvent is practised. In these operations
replacement of that solvent for any reason is often not easy.

If however a mixture of solvents is used, e.g. as in a car paint, then
while recovery of the solvent vapours may be relatively simple,
separation into the individual solvent components is complicated and
costly. Thus, such recovered solvent vapours are generally burnt or
discharged to the atmosphere. Replacement of one or more of the
solvents used in the solvent mixture is usually a simple task in the
event of shortage or rising costs.

It is now time to consider sections of industry which use solvents.

### 3.1. Adhesives
Speed of drying is a major requirement since complete solvent release
is essential to achieve maximum bond strength.

Acetone, methyl ethyl ketone, ethyl acetate, hexane, SBP 2, 3 and
5, high-grade toluene, ethyl and propyl alcohols, methylene dichloride
and 1,1,1-trichlorethane are the main solvents used.

### 3.2. Aerosols
Propellants used in aerosol packs include fluorochlorohydrocarbons,
butane and carbon dioxide. Fluorochlorohydrocarbons, in particular,
often act as solvents for the solid components in such packs; their use,
however, is being curtailed and there is no simple replacement.

### 3.3. Auto Products

Brake fluids need to be high boiling, and di- and tri-ethylene glycol ethers feature in drum and disc-brake fluids. Hydraulic fluids have been formulated on simple alcohol–water mixtures, while radiator fluids are usually based on ethylene glycol. Some closed radiator systems include a portion of lower-cost methyl alcohol.

To prevent carburettor icing after start-up, hydroxyl compounds—ethyl alcohol, isopropanol, diethylene glycol ether and hexylene glycol—have been used successfully for many years. Anti-icing screen fluids are generally based on denatured ethanol or glycols.

Alcohols improve octane rating and have been used successfully in petrol. Brazil, which has very limited indigenous sources of crude oil, uses ethyl alcohol as motor fuel.

### 3.4. Chemical Syntheses

Acetone is the base for a number of other products, including methyl isobutyl ketone and methyl methacrylate monomer.

Benzene, toluene and xylenes are important base chemicals for styrene, phenol, phthalic anhydrides, etc., while alcohols feature in numerous reactions in which substantial quantities of $C_1$ to $C_{10}$ members are used.

### 3.5. Chemical and Pharmaceutical Processes

Among the numerous chemical and similar processes, the following are noteworthy:

#### 3.5.1. Blood Products

By a selective low temperature operation, ethyl alcohol is used to separate various proteins from blood which is no longer suitable for use. The alcohol is recovered.

#### 3.5.2. Glandular Products

Acetone is the extraction solvent commonly used; it, too, is recovered.

#### 3.5.3. Penicillins, Vitamins, Hormones, Alkaloids, etc.

These are among the many other pharmaceutical products processed by solvents in which simple alcohols and ketones feature. In many cases, the solvent is recovered to its former high standard of purity.

### 3.5.4. Oil Products

Methyl ethyl ketone and furfural feature in the production of lubricating oils, while furfural is used in the extractive distillation of $C_4$ hydrocarbons, e.g. butadiene. Sulpholane

is valuable for extracting aromatic hydrocarbons since the operation functions with a very low solvent-to-feed ratio. Sulpholane is also used for removing sulphur impurities from gas streams. Amines, e.g. diethylamine, are used to remove impurities from oils, fats and waxes.

### 3.6. Cosmetics and Toiletries

Ethyl alcohol is the traditional choice for products applied to skin and it features in hair sprays, colognes, lotions, perfumes, deodorants, soap manufacture, etc.

### 3.7. Degreasing, Cleaning

These operations are vital to many major areas of industry. Good solvency for the impurities on metals, plastics, textiles, etc., coupled with their non-flammability ensures a continuing future for chlorinated hydrocarbons. Their relatively low boiling points are well suited to vapour cleaning operations which allow repeated purifying and re-using of the solvent, often at the point of cleansing. The same properties are valuable in liquid and water emulsion cleaning systems.

Methylene chloride, 1,1,1-trichlorethane, perchlorethylene and trichlorethylene feature; the latter has lost some ground in recent years because of its lower threshold limit value.

### 3.8. Fibre Spinning

Dimethyl formamide is used principally in the spinning of acrylic and polyurethane fibres, while acetone is the traditional solvent in the preparation of cellulose-based fibres. The major proportion of polyamide and polyester fibres are melt spun.

### 3.9. Food and Essence Extraction

Solvents used must conform to Statutory Regulations and ethyl and isopropyl alcohols, acetone and hexane feature. End products include vegetable oils, pectin, fish protein and spices.

### 3.10. Herbicides and Pesticides

Apart from their widespread use as chemical intermediates in the preparation of herbicides and pesticides, solvents feature as carriers for the end products. Such solvents need to be high boiling and are mixtures of cyclohexanone or isophorone with aromatic hydrocarbons. White spirit and kerosene are also used to a limited extent.

### 3.11. Printing Inks

Fast evaporating solvents are needed in printing inks applied to impervious surfaces, e.g. wrapping paper and polymer film. Ethyl and isopropyl alcohols, ethyl acetate and toluene are the main solvents used.

In inks which dry by absorption, e.g. newsprint inks, the solvent used is a low cost, low odour, high boiling aliphatic hydrocarbon.

### 3.12. Surface Coatings

This industry uses 35–40% of the available solvents and despite strong competition from solventless systems over several decades, surface coatings are still largely based on solvents. Among the solventless systems worthy of mention are waterborne coatings, 100% solid polymer–monomer systems, powder coatings and polyvinyl chloride–plasticiser dispersions. In the latter, high boiling plasticiser serves as a solvent during application but remains to be an essential component of the final heat-cured coating.

A vast number of resins, vegetable oils and monomers feature in solvent-based coatings often as mixtures and nearly all the solvents listed in Table 1 are used by the surface coating industry.

It is convenient to divide coatings into three sections and the solvents used in each section are given below:

#### 3.12.1. Decorative Drying Oil-Based Coatings

For 50 years, white spirit has been the only solvent used in drying oil-based top, under and primer coats. It is a mixture of paraffin hydrocarbons with 15–20% of an aromatic mixture. The latter confers valuable brushing and flow properties.

#### 3.12.2. Coatings for Industrial Objects, e.g. Cars

This part of the industry is the one that uses nearly all of the solvents shown in Table 1. The coatings are applied in a number of ways including air and electrostatic spraying, dipping and roller coating and

TABLE 2

| Coating | Solvent |
| --- | --- |
| Alkyds—stoving | Xylene, butanol |
| Amino—stoving | Xylene, butanol |
| Cellulose—nitro | Ketones, esters—true solvents |
| | Alcohols—latent solvents |
| | Hydrocarbons—diluents |
| Epoxy | Ketone/xylene mixtures |
| | Glycol ethers |
| Polyacrylates | Ketones, xylene, butanol, glycol |
| | ether acetate |
| Polymethacrylate | Ketones, toluene, glycol ether, acetate |
| Polyurethanes | Esters, ketones, xylene |
| Polyvinyl chloride/acetate | Isophorone, cyclohexanone, aromatics |

in nearly all cases films deposited wet are uneven in appearance and thickness. Solvents provide the means of overcoming such irregularities even in waterborne systems. Often a blend of true, latent and diluent solvents are used in any one coating system. The major coating systems used by industry are now shown along with the solvents in common use (Table 2).

### 3.12.3. Cellophane and Plastic Film
Barrier coatings on food wrapping need safe and quick evaporating solvents. Ethyl and isopropyl alcohols, ethyl acetate and tetrahydrofuran feature, the latter being of particular importance in dissolving highly crystalline polymers of polyvinylidene chloride which have excellent moisture-proofing properties.

## 4. GENERAL FEATURES

Manufacturers of solvents have an excellent record for maintaining high standards of quality from their plants through to the user. It is almost unknown for the most well-known solvents to require any further purification or treatment, even for the most exacting end uses.

Modern analytical techniques have played a major role in identifying and quantifying impurities. Before such techniques were introduced, impurities were usually detected by colorimetric means. The latter are often imprecise as examined by a test for high alcohols in ethanol which is used internationally. In this test, colour reactivity varies from zero for *n*-butyl alcohol to maximum sensitivity for *tert*-butyl alcohol. Once a solvent has been accepted for a process that involves operator contact and/or use by the public, then it becomes difficult to replace that solvent on any grounds except non-availability.

In surface coatings for industrial objects, solvents will maintain their dominant position despite competition from solventless heat-cured coatings. Market penetration of the latter has been hindered by a number of factors including the costs of modifying existing application and drying production lines.

To conclude, the future of solvents derived from petroleum base is assured because of the vast capital invested in manufacturing units and the excellent and consistent quality of the numerous products made. The scope for producing solvents from non-petroleum sources is very limited. Ethyl alcohol is the exception, needing only carbohydrates as the source. The quality of the carbohydrates is, however, all important as there are considerable problems associated with stillage disposal.

# PART II

# APPLIED ASPECTS

# 4

# SOLVENTS USAGE IN PAINTS: NEW DEVELOPMENTS

L. TURNER

*Shell Research, Amsterdam, The Netherlands*

## ABSTRACT

*The surface coatings industry does not change dramatically from year to year. However, it is an industry which, although conservative, has always undergone progressive improvements in the quality of its products and methods of application. A great deal of research has been done and is continuing to be done by surface coatings companies, end-users, machine manufacturers and raw material suppliers. The solvent producers have met the new demands by producing an increasingly wide range of solvents and improving the quality of existing grades. Although solvent free and solvent reduced formulations have been developed the vast majority of surface coating formulations are still based on organic solvents in their many forms.*

*In architectural paints, hydrocarbon fractions of the 'white spirit' type are still the most widely used solvents for gloss and semi-gloss formulations. Recent investments in hydrogenation facilities have made available a new range of solvents with similar evaporation rates to white spirits but with almost zero aromatic content. This manufacturing route has also improved the quality of special boiling point spirits which are important ingredients in paint, adhesive and printing ink formulations.*

*The development of water-soluble binders has led to more attention being given to water-miscible solvents. Traditionally, these solvents have been based on glycol ethers derived from ethylene oxide; increasing interest is now being given to the range of glycol ethers that can be made available from propylene oxide.*

*Another area of development is in low-odour, slow-evaporating solvents for end-uses, such as industrial stoving paints. Diester solvents*

55

*manufactured from a series of dibasic acids are beginning to show promise and give formulators further flexibility in their choice of volatile components.*

## 1. INTRODUCTION

The surface coatings industry is a major user of industrial organic solvents. For instance, the consumption of the West European paint industry is as shown in Table 1. If we examine the types of binders used we immediately see that a tremendous range of organic solvents are needed both from a solvent power and evaporation point of view (Table 2). The end-uses of these paints are shown in Table 3.

TABLE 1
Western Europe—1980 Paint Consumption

| Constituent | $10^6 \times tonnes$ |
|---|---|
| Resins | 1·2 |
| Solvents | 1·7 |
| Pigments | 1·7 |
| Water | 0·5 |
| Total | 5·1 |

TABLE 2
Western Europe—1980—Types of Binder (%)

| Binder | % |
|---|---|
| Alkyds | 44·6 |
| Polyurethane | 1·7 |
| Urea formaldehyde/ melamine formaldehyde | 5·5 |
| Cellulosics | 3·8 |
| Epoxys | 5·1 |
| Phenolics | 3·0 |
| Acrylics | 8·5 |
| Vinyls | 20·8 |
| Others | 6·8 |

TABLE 3
End-use Breakdown

| End-use | % |
| --- | --- |
| Decorative | |
| Trade | 35·6 |
| Retail | 17·5 |
| Transport | |
| First finish | 5·4 |
| Refinish | 6·0 |
| Wood finishes | 8·1 |
| Coil coating | 1·4 |
| Powder coating | 1·2 |
| Other industrial | 20·2 |
| Marine | 3·1 |
| Others | 1·5 |

As decorative paints account for about 50% of the market, I would like to highlight new developments in hydrocarbons for this area and then move on to discuss water-soluble solvents and finally high-boiling diester solvents for industrial stoving paints.

## 2. VERY LOW AROMATIC HYDROCARBONS

The odour of conventional white spirits can be improved by the removal of aromatics which also increasingly provides more customer acceptability.

The composition of white spirits will vary somewhat depending on the crude oil used but a fairly typical composition (% wt) would be:

Aromatics    20
Paraffins    54
Naphthenes   26

Since aromatics have the strongest odour there is a trend to white spirits with lower aromatics content especially in household paints. These solvents are now readily available and their typical properties are shown in Table 4.

The preferred manufacturing route for these solvents is hydrogenation. It is possible to remove aromatics by extraction, but this

TABLE 4

| | Hexane | SBP 140/165 | White spirit | SHELLSOL | | |
|---|---|---|---|---|---|---|
| | | | | D40 | D60 | D70 |
| Density (kg/litre), 15 °C | 0·675 | 0·742 | 0·773 | 0·762 | 0·783 | 0·790 |
| Distillation properties | | | | | | |
| Initial boiling point, °C | 66 | 142 | 161 | 162 | 185 | 194 |
| Final boiling point/dry point, °C | 70 | 160 | 197 | 197 | 219 | 251 |
| Kauri butanol value | 30 | 31 | 37 | 31 | 30 | 29 |
| Aniline point, °C | 65 | 68 | 55 | 69 | 72 | 75 |
| Bromine index, mg Br/100 g | <10 | <20 | | <50 | <50 | <50 |

leads to solvents with inferior solvent power, i.e. lower naphthenic content. For example, if a feedstock containing typically about 20% aromatics were de-aromatised by the two routes the following compositions would result

| | Feedstock | Extraction | Hydrogenation |
|---|---|---|---|
| Aromatics | 19 | 0 | 0 |
| Paraffins | 56 | 68 | 54 |
| Naphthenes | 25 | 32 | 46 |

The removal of aromatics by the hydrogenation technique has other advantages:

(a)  low sulphur content;
(b)  very low olefin content;
(c)  the possibility of making very high naphthenic SBPs.

## 3. PROPYLENE GLYCOL METHYL ETHERS

It is possible to produce a series of glycol ethers based on propylene oxide similar to those that are manufactured from ethylene oxide. The reaction of propylene oxide, however, gives the possibility of the formation of two isomers.

$$CH_3OH + CH_2\overset{O}{\overset{\diagup\diagdown}{-}}\underset{\underset{CH_3}{|}}{CH} \longrightarrow$$

$$CH_3\text{—}O\text{—}CH_2\text{—}\underset{\underset{CH_3}{|}}{CH}\text{—}OH \quad 98\% \text{ Secondary}$$

$$+$$

$$CH_3\text{—}O\text{—}\underset{\underset{CH_3}{|}}{CH}\text{—}CH_2\text{—}OH \quad 2\% \text{ Primary}$$

1-Methoxy-2-propanol

Further reaction with another molecule of propylene oxide gives the dipropylene glycol methyl ether, which will be a mixture of four isomers.

$$CH_3\text{—}O\text{—}CH_2\text{—}\underset{\underset{CH_3}{|}}{CH}\text{—}OH + CH_2\overset{O}{\overset{\diagup\diagdown}{-}}\underset{\underset{CH_3}{|}}{CH}$$

$$\downarrow$$

$$CH_3\text{—}O\text{—}CH_2\text{—}\underset{\underset{CH_3}{|}}{CH}\text{—}O\text{—}CH_2\text{—}\underset{\underset{CH_3}{|}}{CH}\text{—}OH$$

Dipropylene glycol methyl ether

TABLE 5
Comparison of the Physical Properties of the Ethylene Glycol and Propylene Glycol Ethers

| Solvent | Boiling point, °C | Solubility parameter | Water misc. | Evap. rate |
|---|---|---|---|---|
| Methyl oxitol | 124·5 | 10·8 | — | 0·5 |
| Methyl proxitol | 120 | 9·5 | — | 0·7 |
| Ethyl oxitol | 135 | 9·9 | — | 0·35 |
| Ethyl proxitol | 132 | 9·0 | — | 0·5 |
| Methyl dioxitol | 194 | 10·2 | — | 0·02 |
| Methyl diproxitol | 190 | 8·7 | — | 0·04 |

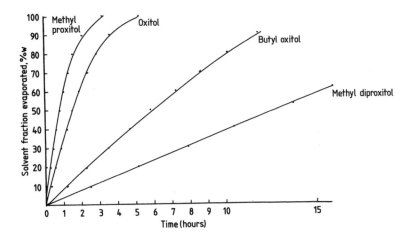

Fig. 1. Evaporation curves of 20% EPIKOTE 1007 solutions in various glycol ethers.

TABLE 6
Reformulation of an Epoxy Coating

| Composition | Formulation (pbw)[a] | | | |
|---|---|---|---|---|
| | 1 | 2 | 3 | 4 |
| EPIKOTE 1007 | 28 | 28 | 28 | 28 |
| Urea formaldehyde | 20 | 20 | | |
| Melamine formaldehyde | | | 16 | 16 |
| Oxitol | 26 | | 28 | |
| Methyl proxitol | | 26 | | 28 |
| Xylene | 26 | 26 | 28 | 28 |
| Viscosity | 150 | 130 | 143 | 133 |
| *After cure 20 min at 205 °C* | | | | |
| Layer thickness | 17 | 16 | 18 | 17 |
| Gloss | 81 | 81 | 78 | 76 |
| Hardness | 149 | 145 | 150 | 149 |
| Ford impact | 75 | 75 | >80 | >80 |

[a] Parts by weight.

A comparison of the physical properties of the ethylene glycol and propylene glycol ethers (Table 5) shows that the propylene glycol ethers are marginally inferior solvents and slightly faster evaporating. A comparison of evaporation curves is given in Fig. 1. If we look at the effect of replacing oxitol with the propylene glycol ether on a simple one for one basis we find no differences in coating performance properties in many cases (Table 6). This could be considered as an exceptionally straightforward case; in other formulations, where the existing solvents are more critical, total reformulation may be necessary.

## 4. DIESTER SOLVENT

This is a relatively new solvent which is a mixture of the dimethyl esters of dibasic acids, the main components being dimethyl succinate, dimethyl glutarate and dimethyl adipate.

$$CH_3-O-\underset{\underset{O}{\|}}{C}-(CH_2)_n-\underset{\underset{O}{\|}}{C}-O-CH_3$$

$$n = 2 \text{ (succinate), } 3 \text{ (glutarate) or } 4 \text{ (adipate)}$$

As can be seen from its properties (Table 7), it is a high boiling, very low coloured, clear liquid with a low pleasant ester-type odour. It is miscible with other organic solvents but only partially miscible with water. It offers a high solvent power and is a relatively low cost, efficient solvent for many resin systems. It is therefore a good solvent for a wide variety of applications in the surface coating, printing ink and adhesive industry or for other applications where a powerful, high

TABLE 7
Properties of Diester Solvent

| | |
|---|---|
| Composition range, % wt | |
| Dimethyl succinate | 20–30 |
| Dimethyl glutarate | 45–65 |
| Dimethyl adipate | 10–25 |
| Content of above esters, % wt | 97 (min.) |
| Colour Pt–Co (Hazen) | 100 (max.) |
| Density at 20 °C, kg/litre | 1·087–1·097 |
| Water, % wt | 0·5 (max.) |
| Acidity as acetic acid, % wt | 0·2 (max.) |

TABLE 8
Comparison of Evaporation Rate of Diester Solvent with
Other High Boiling Solvents

| Solvent | 90% Evap. time (s) |
| --- | --- |
| Diisobutyl carbinol | 13 100 |
| Butyl oxitol acetate | 13 200 |
| Isophorone | 18 400 |
| Dioxitol | 34 600 |
| Diester solvent | 57 000 |
| Hexylene glycol | 66 000 |
| Butyl dioxitol acetate | 300 000 |

boiling solvent is required. It can also be used as an insecticide carrier, as a component for industrial or household cleaning solvent formulations, a chemical reaction medium, and as a plasticising agent.

In terms of evaporation rate, the diester solvent has an intermediate position when compared with other high boiling solvents (Table 8). It can, therefore, be ranked with the faster group which find application as retarders and flow control agents in industrial stoving paints and with the slower group which are often used in emulsion paints. However, in view of its ester structure, it is only recommended for use in emulsion paints with a pH between 6 and 8.

As with any new solvent its introduction into blends will require some reformulation to compensate for the small differences in solubility and volatility characteristics. Suggested areas of application are:

Automotive coatings
    Acrylic lacquers or enamels
    Epoxy primers
    Polyurethanes
Industrial finishes
    Coil coating
    Wire enamels
    Appliance coatings
    Container coatings

# 5

# SOLVENTS IN THE AUTOMOTIVE PAINTING PROCESS: A NECESSARY EVIL?

I. C. MILLER

*Ford Motor Company Ltd, South Ockendon, Essex, UK*

## ABSTRACT

*Although the cost of solvent used in automotive finishing does not form a significant part of the overall manufacturing cost, there is considerable room for improvement. Apart from the actual material costs, the use of volatile organic solvents adds significantly to the cost of equipment installations to meet safety requirements.*

*This chapter attempts to review the research into alternative materials that do not contain organic solvents or those that have a reduced volatile content, which can meet our stringent material requirements. The following areas are discussed and the resulting chemical, manufacturing and economic problems are considered: solid powder coatings; 100% solids, low energy coatings; water-based materials; high solids, solvent-based materials.*

*Existing processes are also discussed in terms of new equipment developments, e.g. the increase in cleaning solvent usage due to the introduction of automatic equipment, creating the need for solvent recovery.*

## 1. INTRODUCTION

It is very easy to imagine, as a specialist paint engineer, that the most important part of automotive manufacture is the finishing process. Unfortunately, this is far from the case, as the total cost of the finishing process represents only a small percentage of the total manufacturing cost of car production. However, the high volume which Ford produce can conceal many inefficiencies which could be improved. I do not intend to give a detailed account of the techniques used in the finishing of car bodies, but rather to emphasise the efforts made to improve the process, within the constraints of a highly competitive market.

In this chapter I will discuss two areas of conflicting interest. On the one hand, the research into new materials to reduce solvent usage and on the other, changes in either market requirements or the introduction of automatic equipment resulting in increased solvent expenditure. I will deal with the various aspects of these 'equal and opposite' forces in turn.

Initially, I would like to explain the reasons for my choice of what appears to be a negative title. In a large competitive business such as ours, any process or material comes under very close scrutiny to discover ways of reducing costs. The overall objective in automotive finishing is to produce a product which satisfies our customers in terms of appearance and corrosion resistance. The process of finishing what is essentially a metal box, comprises a number of separate chemical stages, the majority of which involve the spraying of solvent-based paints. The approximate total solvent usage per plant is 2·4 million litres per annum of mainly xylol. As chemists are in a minority in Ford, the intricacies of paint rheology and behaviour of solvent mixtures, etc., are not generally appreciated. What is noticed, especially by my colleagues in equipment engineering, is that the use of solvent-based products adds significantly to the cost of equipment installations to meet both environmental and safety regulations. Ford, in the USA, is under similar pressure from their environmental legislation to reduce solvent emission. These facts, in conjunction with pressures to reduce operating costs, have caused many engineering managers to ask the question, 'Is solvent necessary or can we avoid using it altogether?'.

If one takes this first question literally, the answer must be: No! There have been many material developments in recent years that avoid the use of volatile solvents altogether or have attempted to reduce the quantity. I would like to briefly describe some of these developments and their potential application in automotive finishing.

## 2. AREAS OF DEVELOPMENT FOR REDUCING SOLVENT USAGE

In general terms, there are four areas of development:

Solid powder coatings

100% Solids, low-energy, coatings.
Water-based materials.
High solids, solvent-based paints.

These will now be discussed.

### 2.1. Powder Coatings

There have been significant developments in 'powder-based' materials in the past few years. The process is based on the dry mixing of components, followed by melt and extrusion, grinding to the correct particle size, and then application by electrostatic spraying. In theory, this type of material is thought to be safer due to the lack of solvent, and can have a material efficiency of approximately 99%. However, a number of problems have barred the way to its widespread use in the automotive industry. The application of this material is totally dependent on the use of 'electrostatic' charging, which causes problems when coating complicated shapes. Although improvements have been achieved, these materials still do not match conventional paint in terms of finish, and rapid colour changing with powder is extremely difficult. In certain instances where these restrictions do not apply this process has been successfully used. In the Ford Motor Company, powder has been used in the finishing of black bumpers, whereas in other car manufacturers, powder coatings have been used in relatively small amounts for various applications. This field requires further development before finding a major application in the finishing of automotives.

### 2.2. Low Energy Materials

A more recent field of research is in the use of 'radiation cured' materials which use low molecular weight monomers and 'reactive' solvents to produce a low-viscosity, 100% solids paint. These materials are cold cured by either ultra-violet rays or a beam of electrons, making this process suitable for plastics. The aim of this development is to reduce the amount of energy used in curing. However, a side-effect is that the resulting materials contain very little volatile solvent. The original development work for one type of electron beam equipment was carried out by the Ford Motor Company in the USA. There are some disadvantages, the most significant being the need for an inert atmosphere during the curing process and the problem of irradiating complicated shapes. Other companies have, more recently, applied this process with more success to the finishing of automotive wheels but this method is not generally applicable to the painting of car bodies.

## 2.3. Water-Based Materials

In the area of 'water-based', or perhaps more truthfully, 'water-thinned' materials, we have been more fortunate. I would stress the latter title because in most instances of industrial paints of this type, water is the major diluent, but some volatile solvent is still present, perhaps as much as 10% by volume when in use. Nevertheless, materials in this category have scored some degree of success in the automotive field, particularly in the case of electrophoretic primers. In principle, this process involves the immersion of the car body in a tank of water-based primer, and a high voltage being applied between the body and the tank. The main advantage of this process lies in the complete coverage of the body, even of hidden areas, and the ability to 'rinse' off the excess paint using demineralised water, from which the paint can be recovered by an ultrafiltration process.

The success of this material is due to the improved corrosion resistance which compensates for some of the considerable on-costs, one of which is the increased energy usage in the curing of water-based materials. Attempts at replacing more conventional paints in primer-surfacer or top-coat applications have not been so successful. This is due to some extent to the material deficiences, but mainly to the extra air conditioning requirements in the spray-booths and longer stoving time. A well-known US automotive manufacturer calculated a three-fold increase in energy costs following an extensive conversion of the total finishing process to 'water-thinned' materials. In the US the 'justification' came from stringent legislation regarding envrionmental pollution. As this is not the case in the UK at the moment, we cannot justify a similar switch, and our own experience of the more conventional sprayable 'water-thinned' materials has been disappointing in terms of appearance.

## 2.4. High Solids, Solvent-Based Materials

This avenue of research is not particularly innovative but is simply an attempt to improve existing solvent-based paints by increasing the solids content. The relevance of this work becomes apparent when it is realised that the average solids content of automotive paints is only 35%, i.e. approximately 65% of every litre of paint Ford uses is solvent which is lost into the air! Trials have been carried out with paints containing up to 60% solids but with varied results. The main effect is a less acceptable appearance of the paint after stoving. This finish may be acceptable for primer-surfacer but not for final top coats. The pressure behind such developments is more dependent on future legislation than a corporate wish to reduce solvent usage.

## 3. CHANGES RESULTING IN INCREASED SOLVENT USAGE

I have discussed briefly Ford's investigations into alternative materials with the aim of reducing solvent usage and solvent emission. I would now like to turn to the other side of the coin, i.e. changes that have inadvertently increased solvent usage. Two of the main changes are:

Changes in sales and marketing requirements.
Improved techniques involving automatic equipment.

These will now be discussed.

### 3.1. Market Trends

As an engineer, I do not feel qualified to comment on the reasons for changes in sales and marketing requirements. In many instances though, these changes result in more work at plant level in terms of the finishing processes. A good example was the adoption of two-coat metallic paints. In the past, the range of metallic paints in regular use in the automotive field were single-coat materials, i.e. a dispersion of aluminium flake in a conventional paint. Five years ago, certain continental manufacturers pioneered a two-coat system, i.e. a base coat containing the colour and metallic content, followed by a top coat of clear lacquer. There is no doubt that the final appearance of this product is far superior; this is confirmed by the general adoption of the two-coat metallic process by the majority of European manufacturers. However, there are disadvantages with this material. First, the base coat has only a 10–15% solids content, or 85% solvent! Secondly, to make matters worse, the solvent mixture used is more expensive, being at least 40% butyl acetate, in addition to the normal xylol (xylene). Although only 30% of the production is painted in metallic colours, this has resulted in a considerable increase in solvent usage, not only for 'cutting' or dilution of the feedstock, but also for cleaning purposes in view of the fact that this process requires twice as many painting stations as the single-coat material. Thus for every body finished in two-coat metallic paint, an additional 2 litres of solvent are used, corresponding to an increase of approximately 1000 litres per day for an average plant.

## 3.2. The Effects of Automatic Equipment

Like many manufacturers in a competitive business, Ford have been forced to decrease operating costs by deploying automation, wherever possible. The painting areas are a particularly good example of this trend. Over the past few years, most painting processes have been 'automated' in some way. Some of the processes introduced are automatic by nature, e.g. the electro-coat tanks mentioned previously. The more conventional paint spray-booths have been automated in a variety of ways. The use of automatic spray-guns, either mounted in a fixed position, or on simple oscillating arms, has become widespread in the finishing process. More recently, painting robots have been introduced to cover the parts not reached by the above-mentioned machines.

Whatever justification is made for this equipment in terms of cost-saving, one of the effects, when automating painting processes, is to increase the solvent usage, particularly when applying colours. To illustrate this, one has to consider the manual method of spraying bodies in colour on a random basis—batch painting not being acceptable in our case. At present we apply 16 colours on a continuous basis which, when manually applied, requires a separate spray-gun for each colour. If automatic machines are used to do this job, we have to use an auto-colour changer for each gun-set, which flushes the previous colour, cleans the gun with solvent and then flushes with a new colour, all within 10 s. To achieve this speed each gun requires approximately 100 ml of high quality cleaning solvent per cycle. During normal production, in one plant alone, this amounts to about 1 million litres of cleaning solvent every year. It is not surprising, therefore, that this increase in solvent use has rekindled an interest in solvent reclamation. The widespread implementation of recycling schemes is dependent on the installation of new equipment designed to collect the solvent used in the cleaning cycle. The maximum potential for collecting and recycling solvent is only about 10–20% of the total usage, the majority of solvent being lost to the atmosphere.

## 4. THE FUTURE

This chapter has described two of the many controlling factors that determine which direction we take in terms of finishing methods and materials. I have simplified the picture; the real situation is far more complex and, significantly, more dynamic. No doubt, 'as I speak', someone is developing new materials of interest to the automotive industry that will affect our thinking.

However, I can say with confidence, that in the short term, the automotive industry will continue to use solvent-based materials, especially for final enamel applications. This view is confirmed by the choice of equipment recently installed in automotive paint shops around the world where the accent has been on new equipment to apply conventional materials. My answer, therefore to any further questions on solvent usage will be that solvent will be 'a necessary evil' for the foreseeable future.

# 6

# SOLVENTS IN INKS AND PRINTING

G. H. HUTCHINSON

*Croda Inks, Edinburgh, UK*

## ABSTRACT

*Solvents continue to be used in quantity as major components of a wide variety of printing inks used for offset lithography, letterpress, flexography, gravure and screen printing processes. Compared with other major components of printing inks—pigments and dyestuffs, synthetic resins, film-forming polymers and modified drying oils, organic solvents have the highest volume and generally the lowest cost.*

*Solvent power and solvent volatility are very important properties influencing the types of inks that can be used for the different printing processes. The use of solvents in modern sheet-fed offset lithographic inks, heatset web offset lithographic inks, flexographic and gravure inks is discussed in relation to the influence of solvent properties on press stability, printability and drying of the inks on the substrate.*

*Formulation of flexographic and gravure inks used for printing food and confectionery packages, requires careful choice of solvents to avoid problems of odour and taint arising from traces of residual solvent in the print, but ink solvent release properties and efficiency of the press-drying equipment are equally important.*

## 1. INTRODUCTION

Two factors have influenced progress in ink technology: the need for inks to print on an increasing variety of new papers, boards, plastic packaging films, metal foils and metal containers, and the demand for inks to print at higher production rates on faster machines. Solvents, the highest volume, lowest cost components of inks, have played an important role in helping to meet these requirements. Factors likely to influence future trends in solvents for inks are the economics of petrochemical-derived materials, the influence of health and safety legislation and environmental pollution problems. In this connection, developments in solvent-free ultra-violet curable inks and water-based inks will be reviewed.

## 2. THE FUNCTION OF SOLVENTS IN PRINTING INKS

Solvents have the function of dissolving synthetic resins, film-forming polymers, modified drying oils and dyestuffs to form vehicles and dye solutions which can then be pigmented to obtain inks. Solvents can also assist in the wetting and dispersion of the pigment.

Evaporation of the solvent component is a major drying mechanism in heatset web offset lithographic printing, flexography, gravure and screen printing processes. Penetration (absorption) of the solvent components into pores of permeable papers and boards is operative in all printing processes but the actual contribution to ink drying depends on the substrate and the printing process. For example, a combination of solvent penetration and oxidation drying is fundamental to the rapid drying of many sheet-fed lithographic inks printed on coated papers. The most important property of the solvent is its power to dissolve vehicle components, followed by the rate of evaporation controlling drying on the substrate and stability on the printing machine.

Effects of solvent on the substrate to be printed are also important. In packaging printing any residual solvent in the dried print could give rise to objectionable odours and taints in food and confectionery products and for this field the use of the least odorous solvents is desirable.

## 3. SOLVENTS IN INKS FOR THE DIFFERENT PRINTING PROCESSES

Figure 1 outlines mechanisms of letterpress, offset lithographic, flexographic and gravure printing processes.

In letterpress and lithographic printing processes the ink is carried to the printing plate by means of a train of rubber rollers which means a

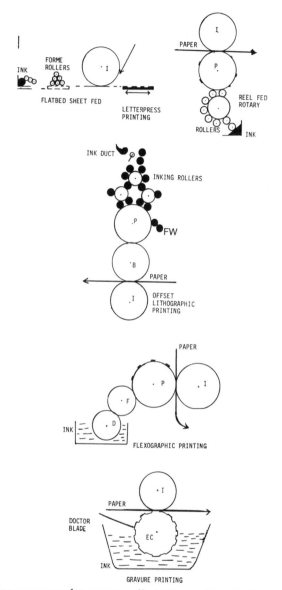

Fig. 1.  Printing processes: letterpress; offset lithographic printing; flexographic printing; gravure printing. I, Impression cylinder; P, plate cylinder; B, blanket cylinder; F, forme roller; D. ductor roller; EC, etched cylinder; FW, fountain water.

thin film of ink is distributed over a large surface area. Both these processes preclude the use of volatile organic solvents that could swell or attack rubber rollers and blankets, or because of high evaporation rates cause premature ink drying or tackiness on the press rollers. Heat development on rollers during press running also accelerates this unwanted evaporation. In letterpress and lithographic printing, the drying mechanisms could be those of penetration (absorption), precipitation, evaporation, autoxidative polymerisation or combinations of these methods. Ultra-violet curable inks dry mainly by photochemical polymerisation.

The ink distribution systems of flexographic and gravure presses are quite different from those of letterpress and lithographic machines. There are no long roller trains, therefore no solvent volatility

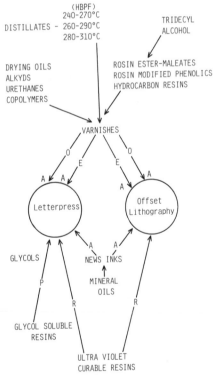

Fig. 2. High boiling solvents and resins. Ink dries by: A, absorption; E, evaporation; O, oxidation; P, precipitation; R, photopolymerisation.

restrictions. In fact, the solvents must be volatile because the drying process depends essentially on solvent evaporation. The range of solvents for flexographic inks is limited to the types which, in high concentration in inks, do not cause swelling or attack on rubber printing stereos, photopolymer relief plates or forme rollers. There are no such limitations in the use of solvents for gravure inks and for this field the ink maker has a wide choice of resin and solvent combinations.

Figures 2 and 3 outline the solvents and resins used in inks for the major printing processes—offset lithography, letterpress, flexography and gravure. Like gravure printing, the screen printing process allows the use of a large range of solvents and resins in the inks. The limiting factor with screen inks is solvent evaporation rate as the printing area is exposed to the atmosphere. The solvent must provide screen stability, yet evaporate quickly from the substrate with minimum energy requirements.

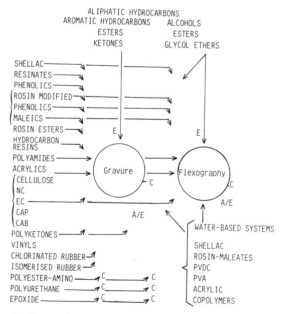

Fig. 3. Low boiling solvents and resins. Inks dry by: A, absorption; C, chemical curing; E, evaporation.

## 4. SOLVENTS IN SHEET-FED OFFSET LITHOGRAPHIC INKS

On coated papers and boards, modern offset lithographic inks dry by a two-phase separation mechanism. Inks are based essentially on pigments dispersed in a vehicle containing, for example, high melting rosin modified phenolic resin and drying oil modified alkyd, dissolved in close-cut high boiling petroleum fractions (HBPF). Solvent boiling ranges are typically 260–290 °C or 280–310 °C, aromatic content is 16 to 20% and the aniline point is 76–90 °C (Kauri butanol values are typically 26–30). Other ink components are slip agents (waxes) and driers.

Figure 4 is a cross-section of printed coated paper showing that on printing impression, solvent molecules are forced into the tiny pores of the paper coating and the resulting rapid increase in viscosity of the ink causes quick 'setting' of the ink film to a touch dry condition. Further penetration of solvent occurs while the printed sheet is in the 'pile' and the second part of the drying involves autoxidative polymerisation of unsaturated drying oil chains in the alkyd component thereby leading to a coherent rub-resistant ink film.

Figure 5 shows how the assisted infra-red heating of a sheet-fed offset lithographic ink accelerates the setting and final drying by increasing solvent penetration and accelerating autoxidative polymerisation of the unsaturated chains in the drying oil or alkyd. Some work was carried out in the author's laboratory to investigate

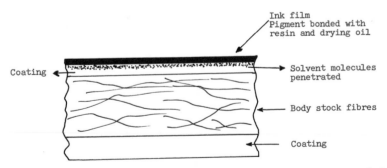

Fig. 4. Ink on paper—cross-section through double-sided coated paper. Quick offset lithographic ink applied to coated paper (immediately after impression). Oxidation 'drying' then completes film formation leading to rub-resistant condition.

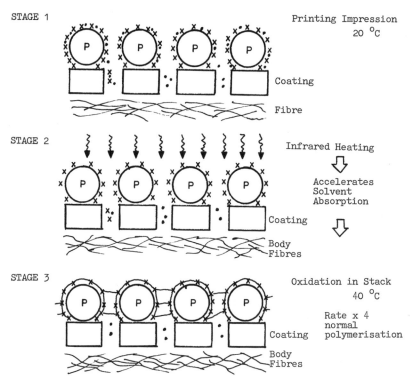

STAGE 1 — Printing Impression 20 °C — Coating — Fibre

STAGE 2 — Infrared Heating ⇩ Accelerates Solvent Absorption — Coating — Body Fibres ⇩

STAGE 3 — Oxidation in Stack 40 °C — Rate x 4 normal polymerisation — Coating — Body Fibres

Fig. 5. Infra-red accelerated drying. P, pigment paticle; ×, resin/drying oil complex; ●, solvent.

whether the rise in temperature over the range 20–40 °C was sufficient to significantly increase the rate of ink 'setting' because of the increased rate of solvent penetration into the paper coating. Table 1 shows the viscosities of typical HBPF solvents at the temperatures indicated. On the assumption that the depth of solvent penetration into the pores of a given paper coating in a given time is inversely proportional to the square root of the viscosity, heating from 20 to 40 °C should bring about an increase in depth of solvent penetration of about 25% over that occurring at 20 °C. No doubt, solvent wetting of the clay particles in the paper coating will be increased by the heating effect but the influence of this on penetration compared to solvent viscosity reduction has not been determined quantitatively.

TABLE 1

Viscosity of Ink Solvents at Various Temperatures

| HBPF solvents boiling range (°C) | Viscosity (centistokes) at temperature (°C) | | | | |
|---|---|---|---|---|---|
| | 20 | 25 | 30 | 35 | 40 |
| 240/270 | 3·64 | 3·23 | 2·91 | 2·60 | 2·40 |
| Low odour 240/270 | 3·23 | 2·97 | 2·65 | 2·39 | 2·18 |
| 260/290 | 4·73 | 4·16 | 3·75 | 3·23 | 3·05 |
| Low odour 260/290 | 4·63 | 4·16 | 3·64 | 3·17 | 3·05 |
| 280/310 | 6·56 | 5·67 | 4·94 | 4·58 | 4·10 |
| Low odour 280/310 | 6·97 | 6·24 | 5·41 | 4·84 | 4·35 |
| Aromatic-free 265/282 | 6·40 | 5·52 | 4·89 | 4·32 | 3·95 |

## 5. HEAT SET WEB OFFSET INKS

Figure 6 is a schematic diagram of a web offset lithographic printing press with drying oven containing gas burners and hot-air knife sections. The inks are essentially pigments, high melting resins, HBPF high boiling solvent combinations. HBPF solvents are typically close-cut fractions, boiling ranges 230–250 °C, 240–270 °C, 260–290 °C, with compositions similar to those of solvents used for sheet-fed lithographic printing inks. Evaporation of the solvent at oven temperatures (hot air), of say 200 to 250 °C, is the principal drying mechanism. The recent trend has been to dry inks at lower oven temperatures (150 to 170 °C) which conserves heat energy and reduces solvent emissions to the atmosphere. The solvents' boiling points cannot be too low otherwise premature evaporation would cause drying of the ink on the roller train. Controlled amounts of tridecyl alcohol are often used along with the HBPF fractions to aid resin solubility, particularly with lower KB (Kauri butanol value) (higher aniline point) solvents. In this way, careful ink formulation can ensure good ink stability on the rollers and fast solvent release in the drying oven. For heatset inks, the use of HBPF solvents substantially free from olefines and aromatics originated in the USA, e.g. to meet the requirements of the Los Angeles Rule 66 legislation. It was found that

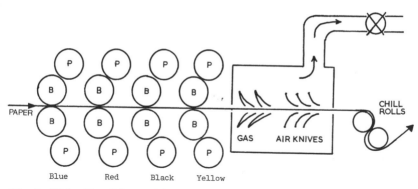

Fig. 6. Web offset lithographic printing press with drying oven containing gas burners and hot-air knife sections. Four-colour heatset web offset inks are used. The web temperature before the chill rollers could be 150 °C for normal drying and <100 °C for low temperature drying. Advantages of low temperature drying include improved ink 'hold-out' and less tendency to blistering of the paper coating. Low emission, high boiling petroleum solvents based on lower aromatic content are used.

the new solvents produced markedly lower odour and gave less visible smoke emission from oven exhaust stacks than the normal types. So-called low emission inks and low temperature drying inks are now used extensively but the low aromatic solvents in the formulations are of course more expensive than normal types.

The solvency power of normal HBPF solvents is retained in a range of high boiling distillates in which hydrogenation reduces the aromatics, olefines and sulphur content and also lessens the odour.

Improvements in high velocity hot-air dryers have contributed to more efficient use of heat energy so that in some installations low temperature drying can be achieved without the need for the formulation of special inks. Even with the low odour grades, the quantity of solvent emitted to the atmosphere can be undesirable in urban areas and it then becomes necessary to incinerate the solvent vapours with an after-burner.

## 6. FLEXOGRAPHIC AND GRAVURE INKS

The solvents used in flexographic inks include water, methylated spirits (ethanol), isopropanol, n-propanol, 2-ethoxy ethanol and esters (ethyl acetate, isopropyl acetate, *n*-propyl acetate). The esters and glycol esters are often used as auxiliary solvents with methylated spirits as principal solvent to keep certain resins, e.g. nitrocellulose, in solution. Low boiling aliphatic hydrocarbons (SBP solvents) and toluene are used with methylated spirits and other low boiling monohydric alcohols with the co-solvent polyamide range of resins. Use of aliphatic and aromatic hydrocarbons and high concentrations of esters can have adverse effects on printing stereos and only small concentrations are tolerable in alcohol inks to be used with photopolymer relief plates. Use of solvent-resistant stereos, however, has widened the range of solvents that can be used in flexographic inks. Large quantities of toluene and SBP solvents are used in publication gravure inks and there are solvent vapour collection and recovery systems at the major plants. Solvents commonly used in gravure inks for packaging are methylated spirits, isopropyl alcohol and *n*-propyl alcohol, 2-ethoxy ethanol, esters, ketones, aliphatic hydrocarbons and toluene. Inks based on methylated spirits dominate this field. They incorporate esters and ketone solvents to accelerate evaporation drying.

The trend in food packaging inks has been away from the use of odorous solvents such as toluene. Ethanol would be less of a problem if retained in traces in the dried ink films. Printers and converters frequently specify maximum permissible limits of total residual solvent $(mg/m^2)$ in the printed substrate as measured by gas liquid chromatography. Information on the type and quantity of individual solvents may also be required. Solvent retention can also adversely affect adhesion of the print to the substrate, especially in the printing of plastic substrates.

The Health and Safety at Work Act 1974 brought about a greater awareness of the potential hazards arising from the handling and use of inks and ink solvents: printers add appropriate solvents to flexographic and gravure concentrates to make 'press-ready' inks. With individual solvents the warning labels are in accordance with 'Packaging and Labelling of Dangerous Substances Regulations 1978' but with inks, lacquers and varnishes, EEC legislation, relating to 'Classification and Packaging of Dangerous Preparations (Solvents)' 73/173 EEC, adapted 82/473 EEC, applies. Here the concentration of a particular solvent, classified as toxic, harmful, irritant or corrosive, determines whether appropriate labelling is needed. Concerning the important ink solvent, 2-ethoxy ethanol, it is appropriate to mention the current debate on possible additional harmful effects reported from animal studies on methoxy ethanol and ethoxy ethanol.[1] It appears that technically similar propylene glycol ethers do not show the same effects and may therefore receive greater usage for inks. It would be premature to exclude the ethylene glycol ethers from use in inks as the studies are as yet incomplete and the question of revised TLVs is still to be resolved.

In the formulation of liquid inks more attention is being given to the use of the solubility parameter concept. Sörensen[2] has described how the viscosity of resin–solvent systems of liquid inks is considerably influenced by intermolecular forces expressed in the solubility parameter concept. Knowledge of the affinity conditions between pigments, resins and solvents, using a 'Resins Solubility Diagram', makes it possible to formulate inks with excellent flow and printability properties. Complex mixtures of solvents in flexographic and gravure inks can give rise to problems during printing due to a shift in solvent balance and it is important for the printer to add make-up solvent to the ink to duplicate the proportion evaporated and not that present in the original ink. Hickman[3] has illustrated this problem with several examples of co-solvent inks. Use of an evaporometer and analysing solvent mixtures at given periods of time can give information on solvent balances. A calculation[3,4] to predict the make-up solvent is based on the concept of activity coefficients.

## 7. FACTORS INFLUENCING THE FUTURE USE OF SOLVENTS

In 1974, shortages and rising prices of petroleum affected the availability of organic solvents and there was an upsurge of interest in water-based inks for flexographic and gravure printing. The subsequent recession brought a return to normal availability of solvents and no pressing needs to evaluate alternative water-based products.

The 1980s are likely to see increasing pressures from legislation on environmental pollution and regulations on safety and handling of inks and auxiliary products. It was a 'smog' problem that led to Los Angeles Rule 66 legislation and stimulated research into solvent-free ultra-violet curable inks, low emission heatset inks and water-based inks. An even greater spur to develop new and improved water-based inks was the more recent legislation imposed by the EPA (Environmental Protection Agency), placing even greater restrictions on the emission of organic solvents to the atmosphere. Although the UK is not likely to experience the problems that led to the Rule 66 legislation there is a provision in the Health and Safety at Work Act 1974 to control certain emissions to the atmosphere. No doubt, there will be increasing pressure from authorities in the UK to ensure cleaner atmospheres in major cities and industrial complexes. All this should lead to a more extensive use of water-based inks for flexographic and gravure printing.

The reduced fire hazard of water-based inks is a big advantage. A further practical advantage of water-based gravure inks and overprint varnishes is their markedly reduced residual print odour in food and confectionery packaging compared with wholly organic solvent-based counterparts. Water-based flexographic inks are well established for printing corrugated cases, paper bags and multi-wall sacks, and the field is expanding with a demand for better quality finish on smoother lined boards and new developments in pre-printing the liner before corrugation. Improvements in water-dispersible polymers in solution, intermediate colloid and emulsion forms have led to improved inks for printing wallpapers and textile transfer papers. Water-based gravure inks for printing football coupons, labels and confectionery cartons have already been commercially realised and a range of water-based overprint varnishes is used to overprint chocolate and confectionery cartons to give absolute minimum print odour. For many applications, water-based flexographic and gravure inks dry extremely well on absorbent papers and boards and this field will grow.

Water-based liquid inks, to print impermeable substrates, such as metal foils and plastic films for packaging, are still largely in the development stages. They have two main shortcomings when compared with solvent-based counterparts. Water requires much more heat to evaporate it by comparison with organic solvents of comparable boiling point (see Table 2), and despite the fact that printed water-based flexographic and gravure inks give a good level of water resistance on papers and boards they yield less resistant prints on plastic films, such as polyethylene, polypropylene and polyester. Their chemical, deep freeze and product resistance properties are also generally inferior to those obtained with the best available organic solvent-based inks. In attempts to improve the drying efficiency of water-based inks on impermeable substrates there has been an increasing interest in radiofrequency heating at microwave and RF dielectric frequencies and in the use of assisted hot air/infra-red heating.

TABLE 2
Properties of Ink Solvents

| Property | Water | Ethanol | Iso-propanol | EGMEE glycol ether | Ethyl acetate | Iso-propyl acetate | n-propyl acetate | Iso-butyl acetate | n-butyl acetate |
|---|---|---|---|---|---|---|---|---|---|
| Boiling point (°C) | 100 | 78 | 82 | 135 | 77 | 88 | 102 | 117 | 126 |
| Latent heat of vaporisation (cal/g) | 540 | 209 | 162 | 234 | 88 | 79 | 80 | 74 | 74 |
| Specific heat, at 20 °C | 1·0 | 0·59 | 0·59 | 0·55 | 0·46 | 0·52 | 0·47 | 0·46 | 0·46 |
| Specific gravity at 20 °C | 1·0 | 0·79 | 0·79 | 0·93 | 0·90 | 0·87 | 0·90 | 0·87 | 0·88 |
| Vapour pressure at 20 °C (mm Hg) | 17·5 | 44·0 | 33·0 | 4·0 | 73 | 45 | 24 | 13 | 8·5 |
| Evaporation rate time (s) ether = 1 | — | 8·3 | 21 | 43 | 3·0 | 4·2 | 6·1 | 15·0 | 12·0 |
| Evaporation rate (volume), BuAc = 1·0 | 1·8 | 3·3 | 2·3 | 0·4 | 6·2 | 5·0 | 2·8 | 1·4 | 1·0 |
| Flash point °C (CC)[a] | — | 13 | 12 | 44 | −4 | 4 | 14 | 18 | 22 |

[a] Closed cup.

Developments in new water-dispersible polymers are enabling improvements to be made in inks for plastic films, but the use of reactive water-soluble crosslinking resins for these applications leads to difficulty in balancing reactivity with press stability. Many water-based flexographic inks contain minor proportions of water-miscible polar solvents—ethanol, isopropanol or glycol ethers, to aid pigment wetting, printability and control the drying speed. A water-based ink, for example, might contain 10–12% by weight of isopropanol but if this concentration is exceeded the benefits of reduced fire hazard may be lost and the 'odour' increased.

For the reasons given above, organic solvents for flexographic and gravure inks will continue to be used in quantity and alcohol-based inks are likely to predominate. Developments in new and improved alcohol-soluble resins would therefore be useful.

## 8. SOLVENT-FREE ULTRA-VIOLET CURABLE INKS

These inks cure instantaneously when exposed in a thin film to a high energy source of ultra-violet radiation. As mentioned previously, they originated in the USA because of Rule 66 requirements. In the UK, the field has grown steadily since its introduction in the early 1970s. Outlets are in offset lithography for general printing of packaging, food and confectionery cartons, labels—pharmaceutical packages and metal decorating, and in letterpress printing for self-adhesive labels. Interest in screen printing is also growing. Instantaneous curing eliminates spray powder in offset lithographic printing and there are no 'solvent' odours. The rub, chemical and product resistances are generally superior to those of conventional lithographic inks. The inks are based on reactive acrylated monomers, oligomers and pre-polymers, photo-initiators and appropriate pigments. Use of roller coating, ultra-violet curable varnishes produces a good gloss and abrasion-resistant finish on record sleeves, for example, which eliminates the need for expensive film lamination. Currently, pigmented ultra-violet inks for flexography and gravure printing have not been commercially exploited due to difficulties in producing a solvent-free pigmented ink of viscosity low enough to print at speed, but clear ultra-violet curable varnishes can be formulated to print by these processes. A commercial disadvantage of ultra-violet curable inks is the relatively high cost of the vehicle components and photoinitiators compared with conventional ink resins, and as there is no solvent as such the inks are significantly higher in cost than conventional inks. Ultra-violet curable inks will be of increasing use in offset lithography and screen printing but they are not likely to make a big impact on flexographic and gravure printing, at least in the near future.

HBPF solvents are likely to continue in use for heatset web offset printing and while there will be growing attention to problems of atmospheric pollution these can be overcome by installation of efficient 'after-burners', if the lower emission types of ink do not meet local demands. The next few years will undoubtedly see economic and technological changes affecting the use of organic solvents but these are likely to be more gradual than dramatic, unless there is a period of shortages coinciding with increased demand when the alternative solvent-free systems will receive more attention.

## REFERENCES

1. The toxicology of ethylene glycol monoalkyl ethers and its relevance to man, ECETOC Technical Report No. 4, July 1982.
2. P. Sörensen, *J. Oil Col. Chem. Ass.*, 1967, **50**, 226.
3. E. P. Hickman, Liquid inks for flexible packaging, Oil & Colour Chemists' Association Scottish Section (Eastern Branch) Printing Inks Symposium, Stirling University, January 1979.
4. Shell Chemicals Technical Bulletin, ICS 70/15.

# 7

# SOLVENTS IN ADHESIVES TECHNOLOGY*

M. J. Welch and R. S. Whitehouse

*Evode Ltd, Stafford, UK*

## ABSTRACT

*Primary solvent selection for an adhesive is a relatively simple matter. Trade and technical literature will quickly define the most useful chemical classes of solvent for a particular polymer or resin. The technical success of solvent-borne adhesives is theoretically explained. Problems associated with important properties of the adhesive, viz. viscosity, drying time, flammability and toxicity are considered. Although the cost per tonne of solvent is low in comparision to solid polymer, it is important to locate the cheapest technically acceptable blend.*

## 1. INTRODUCTION

The basis of adhesion between surfaces is the formation of secondary valence forces. Since it is rarely practical to create surfaces smooth enough and clean enough for such close contact, the links between adherent substances are made by the application of adhesives.

Liquid films can be shown to develop tensile adhesive forces but resistance to shear is lacking unless the liquid solidifies. No doubt readers are familiar with the problem of removing ice from a refrigerator and how simple it becomes after thawing. In reverse, if the secondary valence forces of a high tensile and tough material can be dismantled by dissolution, the material deposited between two

---

* Presented at the 3rd Solvents Symposium in 1980 and revised in 1983.

surfaces, bonds will develop as the secondary valences reassemble on solvent evaporation.

## 2. ROLE OF SOLVENT

Solvents are used in adhesion processes to dissolve solid adhesive film-forming materials and to prepare surfaces for the bonding process. A solution is a homogeneous mixture of two or more substances, the solute and the solvent. One type of adhesive is made by dissolving compounds in organic solvent or water. Adhesive compounds are high molecular weight polymers or materials capable of increasing in molecular weight after the interfacial film has been formed. The solvent is the liquid carrier which provides the prime requisite for all adhesion processes, that of surface wetting, this enabling the adhesive to flow over impermeable surfaces and penetrate into permeable surfaces. As the solvent evaporates, the adhesive film is deposited, and the adhesive strength develops.

## 3. CHOICE OF SOLVENT

Choosing and blending solvents for adhesive application is a proprietary formulating science. Chemistry, mechanics, environment

TABLE 1

| Polymer | Solvent |
|---------|---------|
| Acrylonitrile | Aromatic hydrocarbons, chlorinated hydrocarbons, ketones, esters |
| Natural rubber | Hydrocarbon |
| Butyl and polyisobutylene | Hydrocarbon, chlorinated hydrocarbons |
| SBR[a] (random) | Hydrocarbon |
| SBR[a] (block copolymer) | Hydrocarbon |
| Polychloroprene | Hydrocarbons, chlorinated hydrocarbons, esters, ketones |
| Polyurethane | Ketones |
| Polyvinyl acetal | Esters, alcohol, hydrocarbons, chlorinated hydrocarbons, glycol ethers |
| Acrylic | Hydrocarbons, esters |

[a] Styrene butadiene rubber.

and cost are the factors having the most influence on the selection. Durability in the application environment is the most crucial factor in determining the choice of polymer for the adhesive film which in turn limits the choice of solvent type. The compatibility of polymer and solvent types is given in Table 1.

Solvents, in common use for adhesives, are listed in Table 2.

Solubility parameters are also listed in Table 2; this function is used to help select a solvent or mixture of solvents suitable for a given polymer. For the theoretical derivation of solvent and polymeric solubility parameters refer to Burrell,[1] Hoy[2] and Hansen.[3,4,5] Solvent

TABLE 2

| Solvent | Boiling point or range (°C) | Solubility parameter |
|---|---|---|
| Acetone | 56 | 10·0 |
| Butanol | 118 | 11·4 |
| Butan-2-one | 80 | 9·3 |
| Butyl acetate | 126 | 8·7 |
| Cyclohexane | 81 | 8·2 |
| Cyclohexanone | 156 | 9·9 |
| Diacetone alcohol | 169 | 9·8 |
| 2-Ethoxyethanol | 133·5–136·5 | 9·9 |
| 2-Ethoxyethyl acetate | 145–165 | 8·7 |
| Ethyl acetate | 77 | 9·0 |
| Isobutyl acetate | 117 | 8·3 |
| Isopropanol | 82 | 11·5 |
| Methanol | 65 | 14·5 |
| Methyl isobutyl ketone | 114–117 | 8·4 |
| 2-Methoxyethanol | 123–125·5 | 10·8 |
| 2-Methoxyethyl acetate | 147–152 | 9·2 |
| Methylene dichloride | 41 | 9·7 |
| SBP1 | 46–113 | 7·3 |
| SBP2 | 74–94 | 7·3 |
| SBP2N | 76–92 | 7·3 |
| SBP3 | 100–121 | 7·3 |
| SBP4 | 48–148 | 7·3 |
| SBP6 | 140–163 | 7·3 |
| Tetrahydrofuran | 66 | |
| Toluene | 110 | 8·9 |
| 1,1,1-Trichloroethane | 71–81 | 8·6 |
| White spirit | 153–197 | 7–8 |
| Xylene | 138–140 | 8·8 |

manufacturers have published data which the formulator can refer to; for example Refs 6 and 7. Consideration of Figs 1 and 2 describes briefly how the data are used. Figure 1 shows the envelope of solubility of a hypothetical polymer and a series of solvents from a plot of fractional polarity against solubility parameter. Those solvents falling within the envelope form solutions with the polymer, while those falling outside the envelope will not. Predictions about the effectiveness of solvent blends which include a non-solvent (whose choice may be important for technical or economic reasons) can be seen with reference to Fig. 2. The volume proportion of suitable compositions, for solvents A and B, capable of acting as a solvent is made in the first instance by arithmetically summing the volumes of A and B. A 20% volume proportion of A should just give a solution of the polymer. It would be expected that the solution would improve in clarity and reduce in viscosity as the proportion of B increases. This theoretical approach to formulation is used to limit the number of solvents, their proportions and the number of verification experiments necessary.

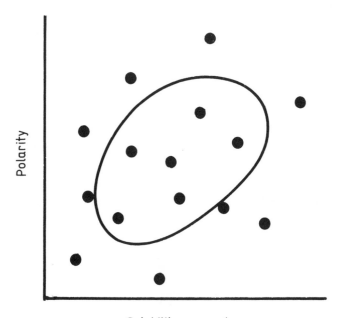

Solubility parameter

Fig. 1. Envelope of solubility of a hypothetical polymer and a series of solvents.

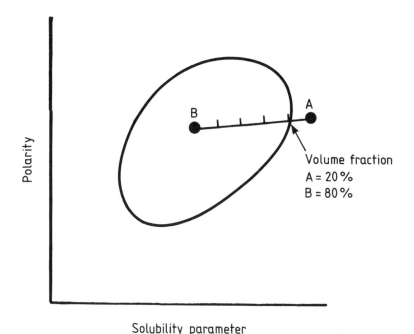

Fig. 2. Solvent A is outside the envelope of solubility but it can be used in a blend with solvent B to form a solvent mixture.

## 4. IMPORTANT ADHESIVE PROPERTIES

The aggregate chemical properties of solvents also influences the selection for the control exercised on one or more of the following properties: viscosity, drying time, flammability and toxicity. These will be considered in turn.

### 4.1. Viscosity

Viscosity is an effective measure of the ability of a solvent, or solvent blend, to break down the secondary valence forces within polymers. It is an important property of adhesives.

The user needs a product with a reproducible viscosity so that his application plant does not require persistent adjustment. The formulator conveniently controls the final viscosity by working to a specification for the solids content of the solution. This is necessary because the structure of the starting materials is disperse and other essential techniques like chemical peptising or mill processing precede the manufacture of the adhesive solution and are subject to variation within specified tolerances.

## 4.2. Drying Time

Drying times, affected by the evaporation of solvent or absorption of solvent and solute, are critical factors in adhesives-use technology, upon which the solvent blender can exercise control from the range of raw materials available. To adhesive users, drying time can have more than one interpretation; it can be defined as the time required before the adhesive films or adhesive film and adherent are ready to be combined (the open time); to others it defines the time when a completed assembly has developed enough strength to be moved or used. It is a matter of striking the balance between surface wetting and the rate of film formation.

An example, in which a formulation for an adhesive—whose major constituents were an acrylonitrile–butadiene copolymer elastomer, a B-stage phenolic resin and chlorinated rubber—was modified for a new application, will demonstrate this point. The original solvent blend consisted of acetone and butan-2-one and in this form the adhesive was used by manual brush application and also as a coating adhesive using a knife spreader to give a dry film thickness of 0·1 mm. It was intended to increase the film thickness and reinforce it with a lightweight fabric reinforcement, both of which created a drying condition causing blisters in the surface.

Blisters are caused by trapped solvent vaporising under a dry surface. When the concentration of solvent at a surface becomes zero before total film formulation is complete, which can happen with solvents of high volatility evaporating rapidly from thick films, diffusion of solvent from the depth of the film to the surface falls to a very low rate. Blisters are the result of vapour forming in small voids within the film. Literature sources gave the following relative evaporation rates for pure solvents: methyl isobutyl ketone (MIBK), 1·4; toluene, 2·3; ethyl acetate, 4·8; butan-2-one, 4·5; acetone, 10·2. By experiment with the formulations in Table 3, using films varying in thickness from 0·1 to 0·25 mm onto glass plates, with drying conditions of 112 °C at low pressure, the formulation containing solvents of lowest evaporation rate at a 3:1 concentration (by volume) was chosen for successful plant production.

TABLE 3
Solvent Blends (Proportions by Volume)

| MIBK | 1 | 1 | 3 | 1 | 3 |
|---|---|---|---|---|---|
| Ethyl acetate | 0 | 1 | 1 | | |
| Toluene | 0 | | | 1 | 1 |

## 4.3. Flammability

The flammability of adhesives based on the combustion energy of solvents is a feature that should not be a hazard if the emission vapours and the environment are the subject of proper controls and regulations. Legal storage and labelling requirements are defined in 'The Highly Flammable Liquids and Liquefied Petroleum Gases Regulations', 1972, and 'The Petroleum (Mixtures) Order', 1929. Adhesive applications should be designed to pay due regard to the emission of solvent vapours and the need to ventilate working places with forced air flow. Using chlorinated hydrocarbons does not reduce the important necessity of providing extraction procedures and enforcing no smoking rules. All the precautions adopted to ensure that adhesives in storage and use are the subject of good standards of industrial hygiene should be applied to adhesive waste. Disposal of waste may be subject to the 'Control of Pollution (Special Waste) Regulations', 1980. The British Adhesive Manufacturers Association has published a valuable compilation of information including the subject of flammability.[8]

## 4.4. Toxicity

Any toxic hazards associated with solvent-borne adhesives stem principally from the solvent fraction. All adhesives should be handled under conditions of good industrial hygiene and all those associated with their use should be aware of the manufacturer's advice that he is legally obliged to provide,[9] to assist in the creation of a safe operation. Solvents can enter the body by inhalation, skin contact and ingestion. All forms of personal contact should be prevented (see Ref. 8, pp. 6 and 7). Solvent vapours in the working atmosphere are assigned 'Threshold Limit Values' (TLVs) by the American Conference of Governmental Industrial Hygienists (ACGIH); the useful relevance of these data has been reviewed by Reed and Scala.[10] The problem of deliberate inhalation has been reviewed by Akerman.[11]

## 5. COST

Adhesive compositions of polymer in solvent usually include between 50 and 85% (by weight) of the volatile constituent. Solvents are not an insignificant part of the raw material costs and add to the total cost by increasing overhead factory costs in other ways:

1. Ventilation and extraction equipment.
2. Firefighting equipment (sprinklers, fireproof doors, extinguishers, etc).
3. Firefighting personnel.
4. Insurance premiums.
5. Transport.

Solvents are used to prepare surfaces for bonding. Intimate surface contact is essential for high strength adhesion and solvents are used for activating, cleansing and degreasing surface preparations. There are adhesive processes that make use of the technique referred to as reactivation. Some adhesive films can be permitted to dry to a tack-free state and after wiping with solvent return to an adhesively aggressive condition. Correctly though, this practice is on the decline; it can be substituted by warming the dry adhesive film (heat reactivation).

## 6. CONCLUSIONS

In conclusion it should be recognised that whilst solvent-borne adhesives retain an important commercial place in the industry, because of their versatility, efforts to find and improve safer and cost-effective alternatives will continue.

Water is an obvious alternative liquid carrier where the longer drying time of water can be tolerated or mitigated. Latex and emulsion-based SBR, polyvinyl acetate, acrylic and styrene acrylic products have been well established for bonding flooring materials, ceramic tiles, acoustic and insulation tiles to surfaces which are porous and receptive to water.

Other forms of adhesive, without solvent, are offered by the industry and include:

(a) Adhesives heated and used molten (hot melts).

(b) Reactive adhesives; the liquid carrier converts to a solid adhesive film.

(c) Filmic adhesives; pressure sensitive, thermally activated and thermally reactive.

The impetus to accelerate the change in use pattern away from solvent-borne adhesives will move slowly. The issues tend to be mutually conflicting. The cost of actual adhesive in a bonded article is small in relation to the total cost and not likely to be dramatically affected by the increasing cost of solvent from a declining oil resource; moreover, the polymeric adhesive material will also increase in cost since most of the associated raw materials used in adhesives are derived from oil. Changing to a different adhesive type will involve some capital appropriation for equipment and development costs to prove new systems. Robotic application is another of today's possibilities; it requires capital funds but is a consideration because there is not an automatic requirement for a different adhesive type and the process can be isolated from human contact. The influence of legislation will probably provide the most impetus and motivation leading to the adoption of new practices.

## ACKNOWLEDGEMENTS

Permission to publish this paper is acknowledged to the Directors of Evode Ltd.

## REFERENCES

1. H. Burrell, *Official Digest*, 1955, **27** (369) 726.
2. K. L. Hoy, *J. Paint Technol.*, 1970, **42** (541) 76.
3. C. M. Hansen, *J. Paint Technol.*, 1967, **39** (505) 104.
4. C. M. Hansen, *J. Paint Technol.*, 1967, **39** (511) 505.
5. C. M. Hansen and K. Skaarup, *J. Paint Technol.*, 1967, **39** (511) 511.
6. 'Solubility Parameters', Shell Industrial Chemicals Technical Bulletin, ICS (X)/75/1.
7. 'Genklene', The Safer Solvent for Formulations, ICI plc Data Sheet.
8. *Safe Handling of Adhesives in Industry* (4th edn), BASA, Southampton, UK, 1983.
9. Health and Safety at Work etc. Act, HMSO, 1974.
10. K. G. Reed and R. A. Scala, *Paint Oil & Colour J.*, June 1971, 776.
11. H. E. Akerman, *Human Toxicol.*, 1982, **1**, 223.

# 8

# SOLVENTS IN THE CHEMICAL SPECIALITIES INDUSTRY*

J. M. C. Roberts and A. V. Bridgwater†

*Johnson Wax Ltd, Egham, Surrey, UK*

## ABSTRACT

*This chapter reviews the purpose, type and extent of solvent use by the chemical specialities industry, discussing and priorising the factors governing solvent selection. The price histories of the major solvents used by the industry are presented, and conclusions are drawn from them, and recommendations for future research and development activity listed.*

## 1. INTRODUCTION

The purpose of this paper will be to review the usage of solvents in the chemical specialities industry, highlighting key materials, and examining how economic factors may affect the future. However, it should be emphasised that, while the analysis regarding the product range, and the type of solvents used is probably valid on an international basis, the review of economic factors is based on UK data only, and can only be considered valid for this area.

## 2. THE CHEMICAL SPECIALITIES INDUSTRY

This branch of the chemical industry is primarily concerned with the manufacture of formulated chemical products, tailored to some specific

* Presented at the 3rd Solvents Symposium in 1980 and revised in 1983.
† Present address: Department of Chemical Engineering, University of Aston in Birmingham, Birmingham, UK.

TABLE 1
Speciality Chemicals

| Business category | Typical products |
|---|---|
| Air care | Air fresheners & sanitising sprays,[a] solid[a] & liquid air fresheners[a] |
| Auto care | Car polishes & cleaners,[a] shampoos, heavy vehicle cleaners,[a] solvent degreasers[a] |
| Biocides | Herbicides,[a] insecticides[a] (mostly for flying & crawling insects) and various plant treatments |
| Carpet care | Carpet shampoos[a] & cleaners[a] in aerosol, liquid & powder form |
| Floor care | Floor care products for cleaning,[a] polishing,[a] including those for manual & machine application |
| Furniture care | Aerosol & liquid furniture polishes,[a] multi-surface polishes[a] & cleaners.[a] Paste waxes[a] & creams[a] |
| Garden care | Specialist garden chemicals such as fertilisers, various plant growth aids |
| Hard surface cleaners | Toilet & washroom cleaners,[a] specialist cleaner products for floors[a] & walls[a] |
| Industrial solvents | Various emulsifiable solvent[a] degreasers & cleaners for industry |
| Laundry care | Fabric softeners, rinse aids[a] & bleaches |
| Metal care | Speciality metal polishes for silver,[a] brass,[a] copper[a] & aluminium[a] |
| Personal care | Products to be used directly on the body, such as hairsprays,[a] shampoos, antiperspirants,[a] deodorants,[a] handsoaps, skin & barrier creams[a] |

[a] Products which include solvents.

end use. The range of products manufactured is quite large, and is summarised in Table 1. Not all of the products presented in Table 1 use solvents, and those that do are marked with an *a* superscript. This industry could not claim to be a major user of solvents in tonnage terms, although the value of the resultant products is significant. Table 2 gives an estimate of the size of this industry in the UK, when measured in both financial and volume terms.

## 3. SOLVENT USAGE

Before discussing products and solvent usage, it is probably helpful to define the term 'solvent' as it will be used in this paper. The solvents

TABLE 2
Chemical Specialities Market Size Parameters

| Business category | UK market size | |
| --- | --- | --- |
| | Volume (thousand tonnes) | Size (£ million) |
| Personal care | | |
| Hair shampoo | 25 | 50 |
| Hair conditioner | 5 | 10 |
| Hair spray | 21 | 42 |
| Antiperspirant/deodorant | 19 | 38 |
| Hand soaps (liquid) | 2·6 | 2 |
| Skin creams | 24·5 | 49 |
| Sub-Total | 97·1 | 191 |
| Garden care | | |
| Fertilisers | 7 | 21 |
| Weed killers, etc., | 4 | 9 |
| Sub-Total | 11 | 30 |
| Floor care | | |
| Floor polish | 13·2 | 13·9 |
| Floor strippers | 5·4 | 2·2 |
| Floor cleaners | 16·5 | 10·0 |
| Floor sealers | 2·8 | 2·2 |
| Sub-Total | 37·9 | 28·3 |
| Furniture care | | |
| Furniture and multi-surface polishes | 11 | 27 |
| Laundry care | | |
| Fabric softeners & spot removers | 33 | 21 |
| Ironing aids | 0·96 | 3 |
| Sub-Total | 34 | 24 |
| Dish washing liquids (hand) | 42·4 | 17·6 |
| Air fresheners | 7·8 | 14 |
| Insecticides | 3 | 10 |
| Auto care | | |
| Car wax, car shampoo, chrome cleaners | 0·5 | 5 |
| Oven cleaners | 2·6 | 4·4 |
| Carpet shampoos | 1·6 | 4 |
| Industrial cleaners | 25 | 3 |
| Metal polishers | 0·1 | 0·5 |
| Grand Total | 274 | 358·8 |

used to manufacture speciality chemicals may be classified into two categories. The first category comprises the refined and processed petroleum distillates, such as kerosine, white spirit and various selected and blended aliphatic and aromatic-containing hydrocarbons. The second category comprises synthetic solvents, for example isopropanol and trichloroethane, most of which are related to oil as a source material, but indirectly, and produced by some form of chemical synthesis. Traditionally, refined and processed petroleum fractions were used wherever possible, owing to their low cost and greater availability. As this paper will attempt to indicate, solvent selection is considerably more complex today.

## 4. CHOICE OF SOLVENT

There are various reasons for incorporating a 'solvent' in a particular product formulation. It might be there primarily as the oil phase in an emulsion polish. In such a case, its solvent power may ultimately be used to remove stains or as a vehicle to take waxes and oils into a substrate, although other solvent materials may also be present in smaller quantities in the formula, to perform this function. A further possibility might be to solubilise a fragrance or active ingredient, and thus render it miscible with a water-based concentrate. Some of the 'use' rationale for chemical speciality solvents are presented in Table 3. Examples of typical solvent use in speciality chemicals are given in Table 4. From Table 4 it is seen that a variety of solvents are used to perform numerous functions, virtually all of which cannot be replaced by non-solvent materials. Sometimes, however, they can be replaced by alternative solvents, and this leads to the question of the method of selection.

## 5. METHOD OF SELECTION

Many speciality chemicals marketed today conform to some basic formula developed quite some time ago. In general, chemical marketers look for a competitive edge, often using increased efficacy or cost per job, as the platform on which to launch or re-launch a product. In many cases, the solvent choice can make a major contribution to success, and this choice occurs during the product

TABLE 3
Solvent Use—Rationale

| Solvent | Typical per cent in product | Reasons for use | Examples |
|---|---|---|---|
| Low boiling, refined and processed petroleum distillates | 15–20% | Oil phase of emulsion | Aerosol furniture polishes |
| Kerosine | 80 | Means of solubilising biocide actives | Aerosol & liquid insecticides |
| Glycol ethers | 1 | To dissolve an ingredient, e.g. resin or perfume, then mix with water-based concentrate | Self-drying, 'drybright' floor polish |
| Methylene chloride | 60 | To solubilise and remove certain oily stains | Dry cleaners & spot removers |
| Ethyl alcohol | 90 | To act as a vehicle to carry polymeric & other materials to a substrate, and by rapid evaporation, lay down deposits | Hairsprays, roller-type anti-perspirants & deodorants |
| White spirit | 70 | To form a paste with non-water-miscible organic materials, e.g. waxes | Paste wax polishes |

development process, which is summarised in Table 5. In this process, the product development brief is usually written by Marketing personnel, and is really a framework within which the product development chemist is expected to work. Specific product attributes tend to be detailed, especially those which are highlighted as being of prime importance, and usually cost constraints are included. A

TABLE 4
Solvent Use in the Chemical Specialities Industry

| Solvent | Typical range (% w/w in formula) | Examples of product applications |
|---|---|---|
| White spirit | 60–90 | Moisture repelling sprays, paste & liquid wax polishes for wood |
| Hydrocarbon with high (30%) aromatics | 65–70 | Industrial degreasers |
| Ethanol | 40–90 | Roller deodorants, hairspray |
| Odourless kerosine (190–250 °C) | 40–80 | Insecticides (flying & crawling insects), aerosols and liquids |
| Industrial methylated spirits | 50–60 | Sanitising sprays |
| $C_{12}^+$ Hydrogenated petroleum distillate (180–220 °C) | 15–90 | Paste wax, metal polishes & laundry aids (pre-spotters) |
| 1,1,1-Trichloroethane | 17–85 | Insecticides, dry clean fabric sprays, industrial cleaners |
| Refined petroleum hydrocarbon (155–270 °C) | 15–50 | Auto polishes |
| Methylene chloride | 20–30 | Insecticides, dry clean fabric sprays, industrial cleaners |
| Isopropanol | 1–45 | Metal polishes, disinfectants, glass cleaners & insect repellants, insecticides (aerosols) |
| Hydrogenated petroleum distillate (140–155 °C) | 10–20 | Furniture & multi-surface polishes, air fresheners, glass cleaners |
| Propylene glycol | 0·5–15 | Roller antiperspirant, metal polishes, air fresheners, toilet bowl & oven cleaners |
| Glycol ethers | 1–5 | Shampoos, hard surface cleaners, floor polish & industrial emulsifiable solvent cleaners |

frequent problem is the balance of technical excellence against the cost of the finished product, and for those products incorporating solvents, it is to the solvent supplier's continued credit that failure to meet a product brief can rarely be put down to technical reasons. That is to say, that the choice of solvents available, usually makes the achievement of the technical objectives relatively easy, but the cost

TABLE 5
Product Development Process

| Step | Activity |
|------|----------|
| 1 | Survey market |
| 2 | Write brief |
| 3 | Preliminary formulation work |
| 4 | Costing exercise on preliminary formula |
| 5 | Modifications on preliminary formula |
| 6 | Performance trials—laboratory based |
| 7 | Consumer research |
| 8 | Evaluation of consumer research |
| 9 | Experimental production runs |
| 10 | Schedule for regular production |

constraints often become the controlling factor. As is hopefully shown in this chapter, this situation is likely to continue for, although alternatives to oil-based solvents are being sought, market forces and energy inflation rates are likely to nullify any major cost savings involved.

## 6. THE MAJOR SOLVENTS USED BY THE CHEMICAL SPECIALITIES INDUSTRY

If Johnson Wax is considered to be a typical example of a major producer of speciality chemicals, thirty-two solvents are used, in the UK, totalling nearly five thousand tonnes per year. Ten of these solvents account for 98% of this consumption, and when considered in quantity terms, therefore, the rest may be considered to be insignificant. The 'top ten' group of solvents judged by the quantities consumed are presented in Table 6, which also tabulates their importance rankings, measured by tonnes of solvent used, tonnes of finished products resulting from that use, and finally, on the total number of products of which a particular solvent is a constituent part. From Table 6, it is seen that the importance of a solvent can vary according to the criteria applied. To the solvent supplier, an important solvent is one which is used by the manufacturers in large quantities, and in our case, such a solvent is odourless kerosine. The user's viewpoint might be different, for important solvents are those which go into important products and these, in turn, are usually products which

TABLE 6

| Solvent | Ranking of importance[a] | | |
|---|---|---|---|
| | Tonnage of solvent | Tonnage of product | Number of products utilising solvent |
| 1. Kerosine | ① | 3 | 5 |
| 2. Hydrogenated petroleum distillate (145–155 °C) | 2 | ① | 2 |
| 3. Trichloroethane | 3 | 4 | 4 |
| 4. White spirit (150–200 °C) | 4 | 8 | 6 |
| 5. Modified white spirit (158–195 °C) | 5 | 10 | 10 |
| 6. $C_{12}^+$ hydrogenated petroleum distillate (180–220 °C) | 6 | 9 | 9 |
| 7. Diethylene glycol monoethyl ether | 7 | 2 | 8 |
| 8. Propylene glycol | 8 | 6 | ① |
| 9. Isopropanol | 9 | 5 | 3 |
| 10. Ethylene glycol monoethyl ether | 10 | 7 | 7 |

[a] Top ranked in importance in each column is ringed.

make a major contribution to the company's profits. That is to say, that even though a relatively small tonnage of a particular solvent may be utilised by a manufacturer, if it goes into his major product, i.e. the one that he both sells the most of and which makes the most profit, then to him it is an important solvent. For the purposes of this paper, the ranking based on tonnes of product produced is regarded as the

TABLE 7

Comparison of Solvent Use by Type—All Other UK Industries vs. Johnson Wax

| | Per cent of total usage (1975) | |
|---|---|---|
| | Other UK industry | Johnson Wax |
| Hydrocarbons | 50 | 72 |
| Oxygenated compounds | 30 | 9 |
| Chlorinated hydrocarbons | 20 | 19 |

important one, for, in general, products which are made and sold in the largest quantities tend to be those that are most successful commercially, thereby satisfying the manufacturer's criteria. On this basis therefore, the solvents presented in Table 6 that occupy the prime positions of importance tend to be those which we have previously categorised as 'synthetics', although petroleum fractions are ranked as numbers 1 and 3.

Table 7 shows how a typical chemical specialities manufacturer in the UK (Johnson Wax) compares with the whole of the industry in the UK[1] with regard to solvent use by type.

## 7. SOLVENT PRICE HISTORY

The cost of solvents is related to the cost of the basic raw material—crude oil—and the cost of conversion—capital cost and operating cost. The increasing real cost of crude oil will have an inevitable effect on many oil-based chemicals, including solvents. The less processing and other value added to an oil-derived solvent, the more closely it will in principle tend to follow the cost of oil. Conversely the more processing and value added to a solvent, the more closely it should follow a representative manufacturing cost index which can be adequately reflected in a plant capital cost index.

This effect is illustrated in Table 8, which compares the prices of the

TABLE 8
Indexed Relative Solvent Costs

| Solvent | Cost indices | |
|---|---|---|
| | 1975 | 1980 |
| White spirit (150–200 °C) | 100 | 100 |
| Kerosine (190–250 °C) | 124·2 | 116·9 |
| Modified white spirit (158–195 °C) | 126·7 | 125·1 |
| Hydrogenated petroleum distillate (145–155 °C) | 166·5 | 143·5 |
| Hydrogenated petroleum distillate (180–220 °C) | 258·4 | 144·4 |
| Isopropanol | 178·9 | 157·5 |
| Propylene glycol | 356·5 | 183·6 |
| Trichloroethane | 328·0 | 210·1 |
| Ethylene glycol monoethyl ether | 385·1 | 217·4 |
| Diethylene glycol monoethyl ether | 386·3 | 255·1 |

previously defined 'top ten' solvents for the years 1975 and 1980. The most basic hydrocarbon fraction, white spirit, is the cheapest solvent and all other solvent costs are indexed to this base for each year using January prices in each case. The next cheapest solvent is a broad-based hydrocarbon fraction with minimal additional processing—odourless kerosine. All the refined and processed petroleum distillates constitute the least costly solvents on this list. As the extent of processing increases, the cost of the solvent increases, as exemplified by the glycol ethers and chlorinated solvents. For comparison purposes, 1975 data are presented alongside the 1980 data, showing how the relative signi-

TABLE 9
Indexed Solvent Price Increases, 1975–1980

| Solvent | Price increase—each solvent assigned base value of 100 (1975) |
|---|---|
| *Synthetics* | |
| Propylene glycol | 132·4 |
| Hydrogenated petroleum distillate (180–220 °C) | 143·8 |
| Trichloroethane | 164·8 |
| Ethylene glycol monoethyl ether | 169·1 |
| Diethylene glycol monoethyl ether | 173·6 |
| *Refined/processed petroleum distillates* | |
| Hydrogenated petroleum distillate (145–155 °C) | 221·5 |
| Isopropanol | 226·4 |
| Kerosine | 242·0 |
| Modified white spirit (158–195 °C) | 254·5 |
| White spirit (150–200 °C) | 257·1 |
| *Economic indicators (1975–1980)* | |
| Chemical plant capital cost index[a] | 192·0 |
| Raw materials and fuel index[b] | 197·5 |
| Crude oil index[c] | 202·4 |
| Retail price index[d] | 212·4 |

*Sources:*
[a] PEI Plant Cost Indexes, Process Economics International, Chemecon Publishing.
[b] Government statistics.
[c] Derived from government statistics on imported oil.
[d] Government statistics.

ficance between the raw material and cost of conversion elements appears to have changed.

It would therefore be expected that the higher value, more extensively processed solvents would not increase in cost at the same rate as lower value, less extensively processed solvents when the cost of crude oil is increasing in real terms. Table 9 shows the indexed price increases from 1975 to 1980 for the 'top ten' solvents, together with some key economic indicators, over the same period. These are the indexed crude oil price increase, the plant capital cost index, the raw materials and fuel index and the retail price index. These data confirm, with only two exceptions, that the refined and processed petroleum distillates have escalated in cost significantly more than the higher value 'synthetic' solvents—the exceptions being a hydrogenated petroleum distillate and isopropanol, which might be explained by 'market forces'. Comparison of these data with the key economic indicators suggests that the indexed prices should lie in the narrow band of 192 (plant capital cost index) to 202 (crude oil price index) or

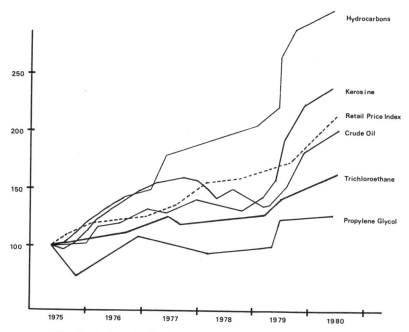

Fig. 1. Indexed price increases and key economic factors.

212 (retail price index) representing the limits of known cost changes over this period. It might be further expected that the Boston learning curve effect[2] would tend to escalate costs at an even lower rate to give a 1980 index value below the suggested minimum of 192.

Actual movements in solvent prices, once again indexed, are shown in Fig. 1 for several important solvents. Again, the significance of the higher price increases of the refined and processed petroleum fractions is evident as well as their price sensitivity to the crude oil price.

## 8. THE FUTURE

Although the chemical specialities industry may not be described as a major purchaser of solvents, this industry needs to carefully monitor and control its costs. The analysis presented above suggests that, wherever technically possible, there may be an economic advantage in moving away from refined and processed petroleum solvents towards synthesised or 'synthetic' solvents as these seem to be escalating in price at a lower rate. It appears very unlikely that any acceptable alternatives to the range of solvents derived either directly or indirectly from oil are going to be available in the short-term, therefore a research and development programme to investigate substitutions might prove rewarding. Economic factors aside, there may be other reasons to look at the more extensive use of synthetic solvents, perhaps dictated by concerns about health, safety and the environment.

The industry will therefore need to maintain a close watching brief on solvent prices, making the type of comparisons that have been outlined in this paper. This is likely to involve much in-house activity, and would benefit from very close liaison with the solvent suppliers themselves.

## REFERENCES

1. Authors' own estimates. See also: A. V. Bridgwater and C. J. Munford, Waste Recycling and Pollution Control Handbook, George Godwin, London, 1979.
2. J. H. Taylor et al., Proc. Econ. Int., 1979, 1 (1) 15–19.

# 9

# HYDROCARBON SOLVENTS IN PESTICIDES: MOVING WITH THE TIMES

K. G. REED

*Essochem Europe, Machelen, Belgium*

## ABSTRACT

*Continuing concern over the environmental effects of using highly toxic chemical pesticides has prompted intensive research into other methods of pest management. While these have met with some success, pests continue to cause immense harm to plant, animal and human health. For the foreseeable future, there is no alternative to the continued use of pesticides in large quantities.*

*Hydrocarbon solvents have long been used in the application of pesticides to plants and animals. Xylene has been most widely used, because of its powerful solvency, ready availability and comparatively low cost.*

*Challenges facing the industry today include the development of new, safer active ingredients and improved methods of application. As a result, a wide range of solvents new to the industry has been made available to assist pesticide manufacturers. These solvents also allow manufacturers to comply with increasingly severe environmental restrictions and can reduce the costs of distribution and application of pesticides.*

## 1. INTRODUCTION

To the average layman, fertilisers are 'good' agrochemicals, pesticides are 'bad'. This attitude to pesticides has been caused by several factors. These include some indiscriminate use of insecticides and herbicides over the years, the well-known persistence of some types of active

TABLE 1
Why Use Pesticides?

| |
| --- |
| 35% OF THE TOTAL VALUE OF GROWING CROPS IS LOST TO PESTS, DISEASES AND WEEDS |
| 120 000 000 HUMAN BEINGS CONTRACT MALARIA EVERY YEAR |
| THE WORLD POPULATION WILL DOUBLE IN 25 (?) YEARS |

ingredients, and a great deal of activity by those concerned with the protection of the environment. Responsible bodies such as the British Agricultural Association have issued reports to inform the public in general of the hazards associated with the use of pesticides, and to explain how these materials may continue to be used without damaging the environment. Other methods of controlling pests are being actively investigated. These include non-chemical methods, of which the development of pest-resistant strains of cereals is an example. They also include novel chemical methods, such as the use of hormones to disrupt the life cycle of selected pests. These new approaches have achieved some significant successes, and hopefully will continue to do so.

Why then, continue to use chemical pesticides? Currently, around two and a half million metric tonnes of pesticides are made each year throughout the world. Even with the use of these staggeringly large quantities, a great amount of harm is done by pests to plant, animal and human health every year. Table 1 shows that 35% of the total value of growing crops is lost to pests, and that malaria, once claimed to be under control, still infects millions every year. Facts such as these, coupled with high rates of population increase in the Third World, imply the continued use of chemical pesticides for the foreseeable future.

## 2. SOLVENTS IN PESTICIDES

In the pesticide industry, as in many others, solvents find many uses, both in the manufacture of active ingredients and in the formulation of pesticides for application. Over the years solvents have been used most widely in formulating emulsifiable concentrates. An emulsifiable concentrate contains a pesticide, together with one or more emulsifiers, dissolved in solvents. Prior to use, the concentrate is diluted with water to form an oil-in-water emulsion, in which form it is sprayed. Traditionally, xylene has been the most widely used solvent, because of its powerful solvency, ready availability, and comparatively low cost. But many other solvents, both hydrocarbon and oxygenated, have been used to meet particular requirements. Solvents continue to be essential ingredients in formulations; examples of the way in which they are helping the pesticide industry to solve today's problems follow.

## 3. MAXIMISING PESTICIDAL ACTIVITY

It has long been known that the manner in which a pesticide is deposited on a target organism can affect its potency. Figure 1, for example, shows some results published in 1971[1] on the influence of the carrier on solutions of herbicides sprayed on to weeds. In this case herbicidal efficiency was significantly improved by the use of an isoparaffinic solvent rather than water.

Currently, pesticide manufacturers are again comparing the activity of pesticides applied via water and via hydrocarbon solvents. 'Flowables' are concentrates of active ingredients dispersed in water which are diluted further with water prior to spraying. They were

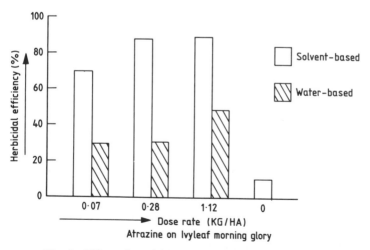

Fig. 1. Effect of pesticide carrier on herbicide activity.

developed to eliminate the dust hazard present when diluting wettable powder pesticides with water before spraying. Flowable concentrates in which water is replaced by hydrocarbon solvents are now being studied by a number of manufacturers. The hydrocarbon-based dispersion is emulsified in water before spraying. This type of formulation is an emulsifiable concentrate of a pesticide insoluble in hydrocarbons. The advantages claimed include improved resistance to wash-off by rain, and reduced volumes for spraying. For this application, mild solvents which disperse rather than dissolve the active ingredients are needed. Either hydrogenated petroleum fractions, in which aromatic components have been converted to naphthenes, or isoparaffinic solvents are recommended.

## 4. MAXIMISING APPLICATION EFFICIENCY

One area of continuing concern is the very low percentage of pesticides applied which are effective in killing pests. Table 2 gives some examples for foliar spray, soil incorporation and aerial spraying. The reasons given for the very low efficiencies are poor transfer to the target organism and volatilisation of the active ingredients.

Improved transfer has been achieved in recent years by electrostatic

TABLE 2
Application Efficiency for Pesticides

| Pesticide | Method of application | Receiving organism | Proportion of applied dose taken up (%) |
|---|---|---|---|
| Dimethoate | Foliar spray | Aphids on field beans | 0·03[a] |
| Lindane | Foliar spray | Capsids on cocoa barley plants (for mildew control) | 0·02[a] |
| Disulfoton | Soil incorporation | Wheat plants (for aphid control) | 2·9[b] |
| Lindane/Dieldrin | Aerial spraying of swarms | Locusts | 6·0[c] |

[a] Estimated by comparing amounts applied with those needed to kill infestation if treated topically.
[b] Measured uptake in pot experiments.
[c] Estimated from analysis of killed locusts.

spraying, made possible because all plants possess a small but definite negative potential. In electrostatic spraying positive charges are generated on droplets of pesticide solutions which are then attracted to plants, rather than hitting them at random. In the UK, ICI has developed the Electrodyn sprayer; using this spray, coverage in cotton has been increased by an average of 250% over that achieved with the same sprayer using uncharged particles.[2] The electrostatic spraying of dispersions and water-in-oil emulsions, as well as solutions, has been described by ICI.[3]

Important parameters in preparing formulations for this type of application are resistivity and viscosity. An ICI patent[4] defines acceptable ranges for formulations as $1 \times 10^6$ to $1 \times 10^{10}$ ohm cm for resistivity and from 5 to 50 centistokes for viscosity. In the formulation, hydrocarbon solvents act as moderators for the other components as well as reducing raw material costs. Table 3 shows how the properties of the hydrocarbon solvents compare with those of the complete formulations, and also shows that the batch-to-batch reproducibility in resistivity, which is an important parameter in formulation design, is acceptably small.

TABLE 3
Hydrocarbon Solvents in Electrostatic Spraying

| Solvent/formulation | | Viscosity (centistokes, 25 °C) | Rate of change of viscosity, 10 °C to 40 °C (centistokes/°C) | Resistivity (ohm cm) |
|---|---|---|---|---|
| Exsol D 60[a] | | 1·50 | 0·02 | $5 \times 10^{14}$ [c] |
| Solvesso 150[b] | | 1·20 | 0·013 | $8 \times 10^{12}$ [d] |
| Permethrin | (50 pbw)[e] | | | |
| Exsol D 60 | (350 pbw) | | | |
| Cyclohexanone | (100 pbw) | 13 to 14 | 0·26 | $2 \times 10^8$ |
| Cereclor 42 | (500 pbw) | | | |
| Permethrin | (50 pbw)[e] | | | |
| Solvesso 150 | (350 pbw) | | | |
| Cyclohexanone | (100 pbw) | 15 to 16 | 0·29 | $6 \times 10^8$ |
| Cereclor 42 | (500 pbw) | | | |

[a] Hydrogenated petroleum fraction.
[b] Aromatic solvent.
[c] Batches vary typically from $2 \times 10^{14}$ to $8 \times 10^{14}$ ohm cm.
[d] Batches vary typically from $5 \times 10^{12}$ to $20 \times 10^{12}$ ohm cm.
[e] Based on examples given in US Patent 4,316,914, February 23, 1982, assigned to ICI.

## 5. MINIMISING PHYTOTOXICITY

Phytotoxicity is the capacity of a material to harm plants. In treating weeds, phytotoxicity is desirable, but in all other instances, phytotoxicity merely damages plants which we are trying to protect.

Solvents used in formulating pesticides are themselves phytotoxic to some extent. Measuring the phytotoxicity of individual solvents does not tell the whole story of their suitability for use in pesticides, since the phytotoxicity of a total formulation is not necessarily the sum of the phytotoxicities of the individual components. Interaction between ingredients, as well as the influence of ingredients on such factors as wetting of and penetration into plants can modify the phytotoxicity ratings of individual components. The type of formulation also influences phytotoxic behaviour. Recent tests by a leading pesticide manufacturer using 1% emulsifiable concentrates of 24 hydrocarbon solvents showed that none was phytotoxic to a range of tropical and temperate plants when applied under hand-spraying conditions. It is expected that differences in phytotoxicity will be observed when the same solvents are sprayed directly on to the plants to simulate the conditions of ultra-low volume application.

To the extent that solvents can be treated in isolation, it is generally accepted that phytotoxicity increases, as shown in Fig. 2, with naphthene content and particularly with aromatic content. Phytotoxicity increases as the volatility of the solvent decreases, presumably because the higher boiling the solvent, the longer it is in contact with the plant before it evaporates. Phytotoxicity should therefore be

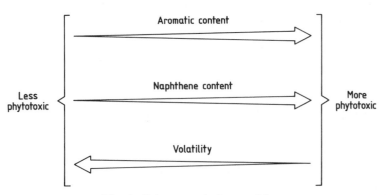

Fig. 2. Solvents and phytotoxicity.

reduced by the use, where possible, of low-aromatic, volatile solvents. Where this is not possible due to the low solubility of the active ingredients, manufacturers are investigating the use of suspensions of pesticides in low-aromatic solvents as an alternative to true solutions as a means of minimising phytotoxicity.

## 6. MINIMISING MAMMALIAN TOXICITY

Pesticides are used because they kill unwanted insects and weeds. Unfortunately, their toxic nature means that they can often harm birds, fish, animals and man. The ideal material is one which is toxic to pests, but non-toxic to other species. Progress in this direction has been achieved by the development of synthetic pyrethroids, man-made pesticides whose structures are related to those of the naturally occurring pyrethrins. Natural pyrethrins, which occur, for example, in some species of chrysanthemum, are very 'safe' pesticides, but the ease with which they break down photochemically makes them economically unsuitable for use in agricultural spraying. The synthetic pyrethroids were developed originally at the Rothampsted Experimental Station, and are now licensed world-wide to pesticide manufacturers. These products are photochemically stable and relatively safe, as shown in Table 4. The ratio 'rats/insects' is a measure of the 'safety' of the active ingredient. Bioresmethrin, one of the synthetic pyrethroids, is the most toxic to insects of the pesticides listed but has by far the highest ratio.

TABLE 4
How Safe are Pesticides? Relative Acute Toxicity of Insecticides to Insects and Mammals

| Insecticide | $LD_{50}$ (mg kg$^{-1}$) to rats (acute oral) | $LD_{50}$ (mg kg$^{-1}$) to insects (topical) | Ratio of rats:insects |
|---|---|---|---|
| Aldicarb | 0·9 | 10 (aphids) | 0·09 |
| Disulfoton | 8·6 | 7·5 (aphids) | 1·1 |
| Parathion | 3–6 | 0·9 (houseflies) | 5 |
| DDT | 118–250 | 10 (houseflies) | 18 |
| Carbaryl | 850 | 4 (mosquitoes) | 212 |
| Dimethoate | 200–300 | 0.7 (houseflies) | 357 |
| Malathion | 1400–1900 | 18 (houseflies) | 91·7 |
| Bioresmethrin | 8600 | 0·2 (houseflies) | 43000 |

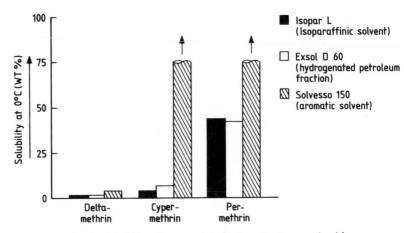

Fig. 3. Solubility characteristics of synthetic pyrethroids.

Solvents are used together with the synthetic pyrethroids, as they are with other pesticides, to prepare solutions, dispersions and emulsifiable concentrates. Figure 3 shows the solubilities of three of the most widely used synthetic pyrethroids in hydrocarbon solvents. Permethrin is sufficiently soluble to be used in solution in any type of hydrocarbon, even the mild, isoparaffinic grades. Cypermethrin can only be used in concentrated solution in aromatic solvents, while Deltamethrin is so insoluble that it cannot be used in solution in hydrocarbons alone, except at very low concentrations.

## 7. COPING WITH LEGISLATION RELATED TO FLASH POINT

The widespread use of xylene in formulating pesticides, because of its high solvency, ready availability and comparatively low price, has already been described. Xylene has been and will continue to be popular with formulators.

Manufacturers of active ingredients, on the other hand, face restrictions with respect to the handling, storage and transportation of solvents having low flash points. These restrictions usually take the form of labelling requirements, but in some cases the use of solvents with flash points below a certain limit is prohibited. In some industries,

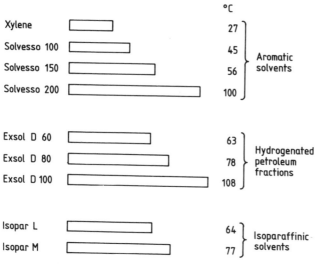

Fig. 4. Flash points of pesticide solvents.

insurance costs when using high flash solvents are less than when using low flash materials.

To assist pesticide manufacturers and formulators, solvents suppliers have developed a range of solvents which varies widely with respect to flash point. Figure 4 shows a typical selection, from which grades can be chosen to comply with or avoid labelling requirements. Undoubtedly legislation is encouraging a move to solvents of high flash points. European manufacturers who market their products in the US, for example, are moving away from xylene to avoid the 'red label' which is mandatory in the US for transportation of products with a flash point below 100 °F (38 °C). In Germany, a law which restricts the storage of more than three tonnes of formulated product if the flash point is below 55 °C has encouraged a move from xylene to Solvesso 150 or equivalent solvents.

## 8. COPING WITH FOOD-RELATED LEGISLATION

The persistence of some pesticides in the environment, together with concerns over the carcinogenicity of some higher boiling aromatic hydrocarbons, has prompted a natural desire to ensure that no health hazards will be created by the use of solvents in pesticides used on food crops. The USA has been most active in establishing requirements for ingredients either to be used in food or to come into contact with food. The Code of Federal Regulations lists the requirements of the Food and Drug Authority (FDA)/Environmental Protection Agency (EPA) for solvents to be used in pesticide applications. While these are strictly applicable only for products sold in the US, they are widely used as guidelines.

Aliphatic solvents, whether straight run, low aromatic or isoparaffinic, have long been classified by the EPA as 'petroleum oils'. As such their use has been relatively unrestricted in formulating pesticides. In the case of aromatic solvents, only xylene has until recently satisfied the requirements of the EPA. Within the last two years a limited number of higher boiling aromatic solvents has been authorised by the EPA for use with pesticides under defined conditions. These are shown in Table 5. The compliance of higher boiling aromatic solvents with the requirements of the EPA is especially important since it complements the move to solvents of higher flash point described previously.

TABLE 5
Aromatic Solvents Exempt from Requirement of an EPA Tolerance

| Solvent | Classified as: | Exempt under: |
|---------|----------------|---------------|
| Xylene | 'Xylene' | 40 CFR 180.1001(C),[a] (D) and (E) |
| Aromatic 100 (Solvesso 100) | 'Xylene range aromatic solvent' | 40 CFR 180.1001(C),[a] (D) and (E) |
| Aromatic 150 (Solvesso 150) | 'Petroleum oil' | 40 CFR 180.1001(B)(3) |
| Aromatic 200 (Solvesso 200) | 'Petroleum oil' | 40 CFR 180.1001(B)(3) |

[a] Meets the requirements of FDA Regulation 21 CFR 172.884(B)(4).
40 CFR 180.1001(B)(3) refers to use on growing crops, in accordance with good agricultural practice.
40 CFR 180.1001(C) refers to use on growing crops or raw agricultural products (stored grain only).
40 CFR 180.1001(D) refers to use on growing crops.
40 CFR 180.1001(E) refers to use on animals.

## 9. CHANGES IN PACKAGING MATERIALS

The choice of packaging materials for solvent-based pesticides is a difficult one. The toxic nature of the formulations usually excludes glass for fear of breakages. Plastics have been used successfully in some instances with water-based formulations, although stress cracking and water permeation are potential problems. Most plastic materials suitable for making bottles are attacked or deformed by organic solvents, so that metal containers, although expensive, have often been used of necessity for solvent-based pesticide formulations.

The appearance within recent years of polyester (PET) bottles which are being used increasingly for packing carbonated drinks, may provide pesticide manufacturers with satisfactory plastic bottles for the first time. Certainly these bottles are extremely resistant to solvent attack. Polyester bottles containing a range of hydrocarbon solvents have been stored continuously at 50 °C for periods up to four months without any visible signs of attack or significant changes in weight. Polyester, like all plastics tested, is permeable to water vapour, but this is not necessarily a problem with solvent-based formulations. Pesticide manufacturers are investigating the use of PET bottles. If these are found to be suitable, a significant advance in packaging pesticides will have been made.

## 10. SUMMARY AND CONCLUSIONS

The pesticide industry, like industry in general, is under considerable pressure to minimise environmental hazards and at the same time needs to reduce costs. Solvents, as traditional components of many pesticide formulations, are contributing significantly to developments in both areas.

From Table 6, which summarises the topics reviewed, the involvement of solvents in pesticides developments can be seen. New, tailor-made solvents are available to help optimise pesticide activity through reformulation, and to improve application efficiency via electrostatic spraying. Solvent selection can help to minimise phytotoxic effects, to minimise restrictions with respect to storage, handling and transportation, and to comply with requirements for food-related applications. Studies have defined suitable solvents for use with the

TABLE 6
Summary—Solvents in Pesticides

| Problems | Possible answer | Solvents contribution |
|---|---|---|
| Increase pesticidal activity | Optimise formulation | Replace water |
| Improve application efficiency | Electrostatic spraying | Controlled, reproducible properties |
| Minimise phytotoxicity | Reformulation | Low aromatic solvents |
| Develop 'safer' pesticides | Synthetic pyrethroids | Solubility studies |
| Reduce labelling, etc. | Reformulation | Higher flash points |
| Food-related use | Complying solvents | Aromatic solvents meet requirements |
| Expensive packaging | Polyester bottles | Storage trials |

newest family of insecticides, the synthetic pyrethroids, and have demonstrated the resistance of polyester bottles to a range of solvents used in formulating pesticides.

## REFERENCES

1. R. J. Burr and G. F. Warren, *Weed Sci.*, 1971, **19** (6) 701–5.
2. N. Morton, *Crop Protect.*, 1981, **1** (1) 27–54.
3. British patent applications 79/16862 (1979) and 78/25660 (1978).
4. US Patent 4,316,914 (1982).

# 10

# THE CHANGING ROLE OF HALOGENATED SOLVENTS IN INDUSTRY

B. P. Whim

*ICI Mond Division, Runcorn, Cheshire, UK*

## ABSTRACT

*The number of halogenated compounds which potentially could be used as solvents is very large but in practice a limited number of $C_1$ and $C_2$ substances are of any significant industrial importance. In the $C_1$ series dichloromethane is widely used for its excellent solvent characteristics. The $C_2$ chlorocarbons—trichloroethylene, 1,1,1-trichloroethane and tetrachloroethylene (perchloroethylene)—are major tonnage products which are in different stages in their growth pattern. The decline in trichloroethylene is due in part to the availability of 1,1,1-trichloroethane which has lower acute and chronic toxicity. The need of the electronics and precision mechanical industries for highly stable, pure cleaning agents has led in the last 15 years to a demand for specialised solvent cleaning systems based on trifluoroethane.*

## 1. INTRODUCTION

The halogenated solvents of industrial importance represent a unique series whose physical properties, especially that of non-flammability, and chemical properties determine their end use. They are a series where changes in application have taken place owing to economic, environmental, toxicological and other pressures over the years. This review highlights some of these changes. The compounds of interest are shown in Table 1.

TABLE 1

| Formula | Common name |
|---------|-------------|
| $CCl_4$ | Carbon tetrachloride |
| $CHCl_3$ | Chloroform |
| $CH_2Cl_2$ | Methylene chloride |
| $CHCl\!\!=\!\!CCl_2$ | Trichloroethylene |
| $CCl_2\!\!=\!\!CCl_2$ | Perchloroethylene |
| $CH_3CCl_3$ | 1,1,1-Trichloroethane |
| $CCl_2FCF_2Cl$ | Trichlorotrifluoroethane (F113) |

## 2. $C_1$ CHLOROCARBONS

### 2.1. Carbon Tetrachloride

This compound is of only minor importance as a solvent and is almost exclusively manufactured as an intermediate for the production of Fluorocarbon 11/12 (95–98·5%). Only about 1·5% is thought to be used in solvents and fire extinguishers. World-wide capacity and production estimates are shown in Table 2.[1] In the 1950s carbon tetrachloride was sold for some solvent applications but its chronic toxic properties preclude its current widespread use for such purposes. Here, indeed, is an example of a change in pattern brought about by an increased awareness of the toxicology, and therefore health risk, from uncontrolled use.

TABLE 2
Estimated Production of Carbon Tetrachloride in 1979 (kilotonnes per year)

|            | USA | Europe | Japan | Rest of World |
|------------|-----|--------|-------|---------------|
| Capacity   | 530 | 724    | 65    | 90            |
| Production | 250 | 468    | 50    | 65            |

### 2.2. Chloroform

Current world capacity has been estimated[1] at 300 kilotonnes per year (Europe, USA and Japan), Table 3 shows world production of chloroform; 90–95% is used as a raw material for F22 ($CHF_2Cl$) production. A small quantity, 2·5%, is used in pharmaceuticals where solvent extraction represents part of this end use. Methylene chloride

TABLE 3
World Prodution of Chloroform

| Year | Production (kilotonnes per year) |
|------|----------------------------------|
| 1960 | 26 |
| 1965 | 54 |
| 1970 | 126 |
| 1975 | 185 |
| 1980 | 262 |

is becoming more important in this industrial sector. Again, although chloroform does have interesting solvent characteristics its toxicity precludes its widespread use in dispersive solvents.

## 2.3. Dichloromethane (Methylene Chloride)

Current world capacity has been estimated[2] at 825 kilotonnes per year; production is shown in Table 4. The principal uses (use patterns) of methylene chloride are shown in Table 5. The industrial uses of methylene chloride are varied and overlap to some extent the hydrocarbon range.

Paint stripping in the industrial sector represents a significant application for this solvent where a need exists to reclaim substandard articles and to restore jigs and spraying equipment. Its powerful solvent action is enhanced by activators in paint stripping formulations and it is applied by a variety of techniques, e.g. dip tanks industrially and in the domestic sector by hand painting type methods.

Methylene chloride is also used in the production of cellulose triacetate from natural celluloses where solvent recovery is a key

TABLE 4
World Production of Methylene Chloride

| Year | Production (kilotonnes per year) |
|------|----------------------------------|
| 1960 | 93 |
| 1965 | 175 |
| 1970 | 331 |
| 1975 | 402 |
| 1980 | 570 |

TABLE 5
Principal Uses of Methylene Chloride

| Application | % |
|---|---|
| Paint stripper (retail and trade) | 25 |
| Acetate film and fibre | 10 |
| Pharmaceuticals | 20 |
| Aerosols | 20 |
| Other (metal cleaning, etc) | 25 |

factor. Similarly in fibre and film production the ability to remove the solvent at low temperatures and the non-flammable properties of the solvent are important.

Its use in the pharmaceutical industry as a process solvent and as an extraction medium for hops, oil seeds and in the decaffeination of coffee also result from its non-flammable, high solvent power characteristics.

A wide variety of other applications exist especially in formulated products such as aerosols where it is used as a vapour pressure modifier, as a solvent for active ingredients and also to suppress flammability. It is also used in adhesives, gasketing compounds and in metal cleaning fluids because of the attractive physico-chemical properties of non-flammability, high solvent power and low toxicity.

The toxicology of methylene chloride has been extensively studied and reviewed in detail elsewhere. A recent CEFIC publication reviews its uses and toxicology.[3]

# 3. $C_2$ CHLOROCARBONS

## 3.1. Trichloroethylene
A recent estimate of the European capacity is 425 kilotonnes and consumption at 250 kilotonnes.[4] The principal uses of trichloroethylene are shown in Table 6. The major solvent application is for metal cleaning in the well-known vapour cleaning process. Limited quantities are used in cold cleaning, extraction and formulations (e.g. adhesives).

Since the major application is in engineering production cleaning the consumption of this solvent is closely tied to the Index of Production. This factor alone has resulted in a decline in consumption of trichloroethylene since the peak of 1970.

TABLE 6
Principal Uses of Trichloroethylene

| Application | % |
|---|---|
| Metal cleaning | 94 |
| Drycleaning and textiles | 2 |
| Extraction | 1 |
| Miscellaneous | 3 |

Many factors however have influenced the pattern of sales of this solvent. Prior to 1955 it was a major solvent used in drycleaning but the advent of dyes soluble in trichloroethylene and the introduction of triacetate fibres led to its replacement by perchloroethylene. In addition the introduction of safer alternatives for cleaning applications, such as 1,1,1-trichloroethane in the mid-1960s, has led to a decline in the consumption of trichloroethylene. Substitution in other parts of the world has also been encouraged; for example, the Los Angeles area of the USA is prone to photochemical smog and the authorities in 1966 legislated against trichloroethylene as a weak photochemical reactant which resulted in its replacement by tetrachloroethylene and 1,1,1-trichloroethane. As stated in the introduction, regional factors of this type can produce unique use patterns for these solvents.

### 3.2. Tetrachloroethylene
Tetrachloroethylene is the major solvent used throughout the world for drycleaning and it is estimated[5] that current West European capacity is 570 kilotonnes. The European market has been recently estimated at 240 kilotonnes in 1979.

Regional differences occur for perchloroethylene and, whereas in the UK it is rarely used for metal cleaning, this application accounts for 30–40% of its use in Western Europe (see Table 7). Prior to 1957,

TABLE 7
Principal Uses of Tetrachloroethylene

| Application | Western Europe (%) | USA (%) |
|---|---|---|
| Drycleaning | 50–60 | 55 |
| Textile processing | 5 | 4 |
| Metal cleaning | 30–40 | 20 |
| Intermediates including captive use | 10 | 15 |
| Miscellaneous | 5 | 5 |

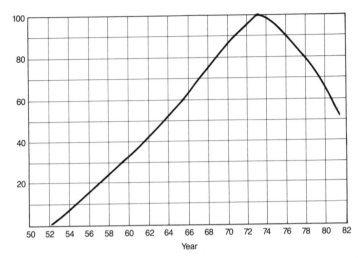

Fig. 1. UK sales of perchloroethylene for 1952–81 related to 1973 value of 100.

white spirit and trichloroethylene were commonly used for drycleaning but both have been replaced in the UK. Figure 1 shows the UK pattern of sales related to sales in 1973 as 100. The consumption of perchloroethylene peaked in 1973 during the period of the coin-operated drycleaning boom and the period of greatest disposable income. The decline since then mirrors greater efficiency in use and improved drycleaning machine design and practice and, of course, the trend to garments which can be laundered at home.

### 3.3. 1,1,1-Trichloroethane
In the UK this product was introduced in the 1960s as a solvent of low acute and chronic toxicity. The current European capacity is estimated at 300 kilotonnes with an estimated European market of 150 kilotonnes.

The physico-chemical properties and the low acute and chronic toxicity of this solvent have contributed to both its overall growth and its broad spread of applications (see Table 8). The use as a production cleaning solvent is the major application and growth in this sector and is one factor in the previously mentioned decline in trichloroethylene consumption. The use of 1,1,1-trichloroethane as a solvent is still in the growth phase.

TABLE 8
Principal Uses for 1,1,1-Trichloroethane

| Application | % |
|---|---|
| Cold cleaning | 30 |
| Vapour cleaning | 35 |
| Formulations, e.g. adhesives, aerosols and inks | 35 |

### 3.4. Trichlorotrifluoroethane

1,1,2-Trichloro-1,2,2-trifluoroethane is the only member of the fluorinated hydrocarbons to be used as a solvent. It was introduced into the UK in the late 1960s for production cleaning of assembled printed circuit boards (PCBs). Electronic production depends extensively on the use of PCBs which, following fluxing and soldering, need to be cleaned to remove residues in order to improve visual appearance and to remove corrosive residues. The presence on the PCB of components, paints, rubbers, etc., which traditional solvents may damage has led to the growth of a market for very specific solvent cleaning processes based on F113. Solvent blends or azeotropes based on this compound enable high standards of cleaning to be carried out without damage to substrates. Such solvent processes have become an essential part of many production schedules and have led to a substantial growth in the use of this solvent.

This solvent is also used in drycleaning where the design of economic machinery has allowed this volatile solvent to be used economically not only for cleaning of suede, leather and special garments but for general drycleaning purposes.

## 4. STATE OF MARKET MATURITY AND FUTURE PATTERNS

The state of maturity of the market for each solvent is dependent on many factors. Clearly the utility of a given solvent for a particular duty is important but economic, toxicological and environmental factors have and will continue to play an important role in solvent choice and therefore growth.

The solvents will continue to be important industrially and attention will turn towards process design to ensure safe and economic use. In the solvent cleaning field in either engineering or drycleaning, for example, the better companies will match their engineering design with correct solvent choice to produce a cleaning process. In this way, this important solvent range will continue to play a major industrial role.

## REFERENCES

1. P. R. Edwards, I. Campbell and G. S. Milne, *Chem. Ind.*, 18 Sept. 1982, 714–18.
2. P. R. Edwards, I. Campbell and G. S. Milne, *Chem. Ind.*, 4 Sept. 1982, 619–22.
3. European Chemical Industry Association, *Methylene Chloride Use in Industrial Applications*, CEFIC-BIT, March 1983.
4. European Chemical Industry Association, *Trichloroethylene in Metal Cleaning and Other Industrial Applications*, CEFIC-BIT, June 1980.
5. European Chemical Industry Association, *Perchloroethylene in Drycleaning, Textiles and Industrial Applications*, CEFIC-BIT, October 1981.
6. European Chemical Industry Association, *1,1,1-Trichloroethane in Metal Cleaning and Other Industrial Applications*, CEFIC-BIT, May 1982.

# 11

# MOISTURE PROBLEMS IN SOLVENTS

G. J. KAKABADSE

*Department of Chemistry, UMIST, Manchester, UK*

## ABSTRACT

*The presence of moisture in organic solvents can create serious problems in the manufacture of industrial products, e.g. paints, inks and polymers, in textile finishing. On the other hand, it is beneficial in the production of polyurethane foams: water being added, sometimes up to 20%, to the polyol. It is shown how corrosion and microbial biodeterioration can be influenced by the presence of residual water in solvents. The removal of water from solvents, usually a costly process, is rendered more difficult by solvent hygroscopicity. The critical role of accurate on-line monitoring of moisture in solvents is discussed.*

## 1. INTRODUCTION

Many industrial solvents, the more important of which are listed in Table 1,[1] commonly suffer from moisture problems. When dealing with the latter it is helpful to consider the following solvent parameters.

(i) *Water miscibility:* Miscible with water in all proportions, e.g. ethanol, acetone (propanone); partially miscible, e.g. methyl ethyl ketone (butan-2-one); practically immiscible, e.g. per-chloroethylene, xylene.

(ii) *Solvent hygroscopicity:* Bonding of water to solvent through hydrogen bonding, viz. bond between hydrogen (already bonded to an atom) and the lone pair of electrons on a strongly electronegative element, such as F, O or N.

129

TABLE 1
US Consumption—1974

| Solvent | Total (×10³ tons) | Solvent use | |
|---|---|---|---|
| | | (per cent) | weight (×10³ tons) |
| Methyl ethyl ketone | 238 | 100 | 238 |
| Methyl isobutyl ketone | 86 | 100 | 86 |
| Ethyl acetate | 70 | 100 | 70 |
| Isopropyl acetate | 30 | 100 | 30 |
| n-Butyl acetate | 32 | 100 | 32 |
| Methylene chloride | 236 | 100 | 236 |
| Trichloroethylene | 175 | 90 | 157 |
| Perchloroethylene | 332 | 88 | 292 |
| 1,1,1-Trichloroethane | 236 | 77 | 182 |
| Diethyl ether | 29 | 65 | 19 |
| Ethyl alcohol (synthetic) | 730 | 56 | 410 |
| Isopropyl alcohol | 855 | 50 | 427 |
| Glycol ethers | 240 | 47 | 113 |
| Acetone | 880 | 37 | 325 |
| n-Butanol | 157 | 20 | 31 |
| Toluene | 2 860 | 20 | 572 |
| o-Dichlorobenzene | 32 | 20 | 6 |
| Xylenes | 2 650 | 19 | 504 |
| Chloroform | 132 | 15 | 198 |
| Chlorobenzene | 172 | 13 | 22 |
| Methanol | 2 700 | 10 | 270 |
| Cyclohexane | 755 | 3 | 22 |
| | | Total | 4 242 |

(iii) *Hardness/softness of solvents.* In this classification,[2] hard solvents (e.g. water) will strongly solvate hard bases (e.g. $F^-$, $OH^-$), whereas solvents with no acidic hydrogen atoms (e.g. acetone) will usually be softer and will have preference for solvating large anions which function as soft bases. Solvents such as benzene are the softest of all.

## 2. EXAMPLES OF MOISTURE PROBLEMS

(i) Reactivity of water towards typical organic bonds can be exemplified by carbanion formation in a variety of solvents, e.g.

dimethylsulphoxide, tetrahydrofuran, diethyl ether, by the reaction:

$$R_3C—H \underset{\text{water}}{\overset{\text{base}}{\rightleftharpoons}} R_3C^-$$

(ii) In polyurethane production, reactivity of water can lead to 'blocking' of isocyanate groups by water,[3] i.e. stopping chain growth:

$$\text{\textasciitilde}R—NCO + H_2O \longrightarrow \text{\textasciitilde}R—NH_2 + CO_2$$

When diluting isocyanate resins with solvents (e.g. xylene, butyl acetate), the water content of solvents is very critical and must not exceed 0·02% $H_2O$. The requirements are even more stringent when selecting suitable solvents for urethane coatings[4] to ensure storage stability, good drying characteristics and blister-free films. The solvents employed in coating systems must, as far as practicable, be free from water.

(iii) In contrast, water in solvents can have a beneficial effect in the production of polyurethane foams.[5] Isocyanate-terminated pre-polymer and excess diisocyanate can be foamed by reaction with water:

$$OCN—R—NHCOO\text{\textasciitilde}OCONH—R—NCO + 2R(NCO)_2 \xrightarrow{H_2O}$$
$$\text{\textasciitilde}OCONH—R—NHCONH—R—NHCONH—R—$$
$$NHCONH—R\text{\textasciitilde} + CO_2$$

The foam density can be controlled by regulating the water content of the polyol over the range 1–20% $H_2O$.

(iv) Accurate control of the water content in secondary butanol is imperative in the catalytic conversion of this solvent to methyl ethyl ketone, moisture being detrimental to catalysts (e.g. ZnO, Cu) by poisoning active sites on the catalyst. It must be borne in mind that raw secondary-butanol can contain up to 25% water.

## 3. EXAMPLES OF MOISTURE PROBLEMS FROM INDUSTRY

The following communications from industry dealing with moisture problems in solvents are gratefully acknowledged.

(i) In paints, moisture can cause instability and 'gassing' of metallic paints, reduced curing efficiency of 2-pack urethanes, gelation of polyvinyl butyral/phosphoric acid etch primers and low electrical resistivity which may give problems with electrostatic spray paints.

(ii) In organic pigments, moisture content has a pronounced effect on the dispersibility of pigments in paints and offset printing inks. Pigments from which water is displaced by solvents, e.g. dry methanol, are most readily dispersable.

(iii) In ceramic inks, residual water in solvents can cause thixotropy. In clays, variation in moisture content can affect the gel strength of organophilic clays widely used as thixotropes.

(iv) Water-immiscible solvents, such as perchloroethylene and xylene, can also cause moisture problems owing to a finite, temperature-dependent solubility of water in them. In cloth finishing, considerable losses can occur when moisture retained in hot perchloroethylene separates on cooling, providing a layer of water on top of supposedly pure solvent. The water frequently cannot be seen and gets pumped across to a working area as pure perchloroethylene which can be very damaging to piece goods as cloth is processed in a nonaqueous medium.

(v) A similar observation has been made in the case of xylene—moisture dissolved in xylene dropping out of solution at low temperatures.

(vi) Residual water in solvents can cause corrosion and flocculation of pigments, particularly $TiO_2$, in solvent-based ink systems.

## 4. CORROSION

Corrosion reactions in organic solvents[6ab] are becoming a frequent occurrence in industrial processes; of special importance are (i) the nature of organic solvent (hard solvents, e.g. methanol, ethanol, being more aggressive than soft solvents, e.g. acetone), (ii) the presence of trace quantities of water, (iii) organic/inorganic acids, (iv) aggressive ions. Points (ii)–(iv) may result from inherent impurity, contamination and solvent degradation.

Aspects of solvent-induced corrosion are discussed by S. Hewerdine in Chapter 12.

## 5. MICROBIAL BIODETERIORATION

Microbial biodeterioration deserves special attention.[7] The presence of moisture in fuel (including aviation fuels), lubricating and hydraulic oils can lead to biodeterioration brought about by microorganisms. While microorganisms can feed on hydrocarbon fuel, *the prerequisite for microbial growth is the presence of water*. Initially, water is present as a contaminant, but once established, microbial growth results in production of more water, since water is an end-product of hydrocarbon oxidation.

The consequences of microbial growth are, first, corrosion, caused by aggressive chemicals produced during microbial metabolism (e.g. acids, sulphide ion) and, secondly, blocking of pipelines, filters and drain holes due to the physical presence of the microorganisms.

## 6. NEED FOR ACCURATE CONTROL OF RESIDUAL WATER IN SOLVENTS

In many industrial processes the water content of organic solvents is an important factor in determining the quality of the final product. This is repeatedly stressed in the following excerpts from feedback from industry, which are gratefully acknowledged.

In solvent recovery, a successful on-line moisture monitor linked to a control system would be of enormous help in the efficient and economical operation of continuous distillation columns. In the manufacture of industrial chemicals, replacing batch methods by on-line analytical control could lead to cost savings by reducing operator effort, thus giving considerable savings in bulk transportation. In the cosmetic/pharmaceutical industry, improved monitoring of methyl ethyl ketone for water could prevent appreciable financial loss. In cloth finishing, losses occur due to inaccurate water determination in the processing solvent. In the distillers' industry, continuous monitoring of ethanol and methanol in ethanol/water and methanol/water mixtures, respectively, would be invaluable in speeding up strength determinations of ethanol coming from the rectifying columns, the strength of which is continually changing, and also in determining when complete mixing of methanol/water had taken place.

Accurate, preferably on-line, measurement of moisture in solvents is therefore highly desirable and would provide cost-effective control of a process.

## 6.1. Currently Used Analytical Methods

Several methods are being employed in industry for the determination of water in solvents, including the following: infrared, Karl Fisher, gas chromatographic, density measurement of distillate, thermometric, nmr, colorimetric, mass-spectrometric, nuclear method, measurement of dielectric constant, utilisation of stabilised RF signal. The main disadvantage with most of these is that they are batch techniques while industry prefers continuous methods.

Although infrared spectroscopy is well established for the continuous determination of water in solvents it lacks general applicability over a wide range of water/solvent(s) concentration without laborious calibration.

## 7. ION-ISOCONCENTRATION TECHNIQUE: A NEW POTENTIOMETRIC METHOD FOR THE DETERMINATION OF WATER IN ORGANIC SOLVENTS

The ion-isoconcentration technique (IICT), developed at UMIST,[8] provides a new rapid method for the determination by direct potentiometry of residual water in organic solvents (Table 2) and individual organic solvents in simple mixtures (Table 3), using ion-selective electrodes which are readily adaptable to automation.

Under IIC conditions, potentials of cells vary systematically and reproducibly with variation in water and solvent concentration, respectively. In the case of a pH glass electrode, the variation in cell potential is proportional, with a high degree of sensitivity, to the

TABLE 2

Accuracy of Water Determination in Solvents under Proton Iso-concentration Condition

| | Water (%, m/m) | | | | |
|---|---|---|---|---|---|
| | *Propanone* | *Methanol* | *Ethanol* | *Propan-1-ol* | *Propan-2-ol* |
| Expected | 0·072 | 0·124 | 0·138 | 0·116 | 0·135 |
| Found | 0·070 | 0·122 | 0·142 | 0·109 | 0·131 |

TABLE 3
Determination of Ethanol in Spirits, (a) Under Fluoride
Isoconcentration Condition (FICT) and (b) by Density
Measurement of Distillate (DD) adopted by HM Customs
and Excise

| | Ethanol (%, v/v) | | | | |
| | Vodka | Whisky | Cognac | Gin | Ouzo |
|---|---|---|---|---|---|
| FICT | 38·1 | 40·0 | 40·5 | 40·3 | 46·2 |
| DD | 37·8 | 40·2 | 40·2 | 40·0 | 46·3 |

variation in the moisture content of organic solvents (in acid solution) at low water content (Fig. 1).

In the system, $10^{-2}$M HCl–water (0–100% m/m)–propanone, shown in Fig. 1, the observed change in cell potential was 353 mV, of which 103 mV occurred over the narrow concentration range of 0–2·4% water.[8] This 'hypersensitivity' of potential at low water concentration, also observed for other organic solvents,[8] forms the basis for the potentiometric determination of residual water in organic solvents by the ion-isoconcentration technique.

Greater accuracy and higher stability of potential have been achieved in cells without liquid junction when using two ion-selective electrodes, e.g. pH glass electrode and solid-state chloride electrode.

As can be seen from the results shown in Tables 2 and 3, IICT can cope with narrow and wide ranges of organic solvent/water concentration.

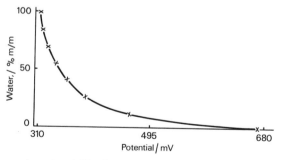

Fig. 1. Change in potential in the system, water–propanone–$10^{-2}$M HCl, using pH glass and saturated calomel (LiCl) electrodes.

Organic solvents (OS) which have been studied in OS/water mixtures include the following: methanol; ethanol; propan-1-ol; propan-2-ol; 2-methylpropan-2-ol; ethan-1,2-diol; ethanoic acid; ethanal; propanone; 1,4-dioxan; formamide; acetonitrile; dimethyl-sulphoxide; glycerol and sulpholane. Recent experiments have included the determination of residual water in aprotic solvents, e.g. toluene, and in oils, the conductance of which was increased by dilution with a known amount of methanol or ethanol.

## 8. CONCLUSIONS

In many industrial reactions the water content of organic solvents is an important factor in determining the quality of the final product. Accurate on-line measurement of water in solvents is therefore highly desirable and can provide cost-effective control of a process.

The ion-isoconcentration technique which has proved satisfactory for the determination of residual water in organic solvents for batch analyses is being currently adapted at UMIST to continuous flow systems; the work is being supported by a Wolfson Foundation Award.

## ACKNOWLEDGMENTS

Assistance by the following is gratefully acknowledged: Dr A. K. Covington (University of Newcastle), Dr I. C. Hamilton (Footscray Institute of Technology, Australia), Dr M. R. O. Karim (Salahaddin University, Iraq), and by my colleagues at UMIST: Miss Nada Al-Yawer, Dr J. Lee, Dr R. Perry, Dr J. Rawcliffe and Dr A. E. Tipping.

## REFERENCES

1. T. M. D. Ball, in *Solvents—The Neglected Parameter* (ed. by G. J. Kakabadse), UMIST Publication, 1977, p. 2.
2. R. G. Pearson, in *Hard and Soft Acids and Bases* (ed. by R. G. Pearson), Dowden, Hutchinson and Ross, 1973, pp. 59 and 70.
3. J. H. Saunders and K. C. Frisch, *Polyurethanes*, Interscience Publishers, 1967, p. 334.
4. Reference 3, p. 532.

5. Reference 3, p. 315.
6. (a) L. L. Shreir (Ed.), *Corrosion*, Vol. 1, Newnes–Butterworths, 1979, p. 1:18; (b) E. Heitz, Corrosion of metals in organic solvents, in: *Advances in Corrosion Science and Technology* (ed. by M. G. Fontana and R. W. Staele), Plenum Press, New York, 1974, Vol. 4, p. 149.
7. J. D. A. Miller, in *Microbial Biodeterioration* (ed. by A. H. Rose), Academic Press, 1981, p. 150; C. Genner and E. C. Hill, *ibid.*, p. 260.
8. G. J. Kakabadse, *Ion-Selective Electrode Rev.*, 1981, **3**, 127; *Anal. Proc.*, 1981, **18**, 225; *Analyst*, 1978, **103**, 1046; *Proc. Analyt. Div. Chem. Soc.*, 1975, **12**, 83.

# 12

## MATERIALS OF CONSTRUCTION FOR HANDLING SOLVENTS IN CHEMICAL PLANT*

S. HEWERDINE

*ICI Engineering Department, Billingham, UK*

### ABSTRACT

*This chapter covers the procedures which should be followed and the factors that should be taken into account in the selection of suitable materials of construction for use in the manufacture and processing of solvents, with particular emphasis on corrosion performance, including a look at some of the likely problems associated with the use of the more common materials of construction.*

## 1. MATERIALS SELECTION

In specifying materials of construction for any application, there are three important steps which need to be followed:

1.1. Listing the requirements to be met by the material.
1.2. Selecting the evaluating candidate materials which are likely to be able to meet these requirements.
1.3. Choosing from these the most economic materials to serve the purpose.

Provided that the first two steps are fully considered, then the third stage becomes relatively straightforward. Detailed consideration is therefore given to steps 1 and 2.

---

* Presented at the 3rd Solvents Symposium in 1980 and revised in 1983.

## 1.1. Requirements to be Met

### 1.1.1. Properties

*1.1.1.1. Mechanical.* This is the strength, toughness and ductility of the material; if these are inadequate, as is the case with many non-metallic materials, then the use of the material as a liner for a material with more suitable mechanical properties (often steel), or with some form of reinforcement, e.g. glassfibre, may be necessary. Other mechanical properties which must be taken into account are fatigue and creep resistance, and flexibility or rigidity. Obviously, the particular requirements will depend on the type of item being considered; for example, materials for a valve seat and a centrifuge basket will require vastly different mechanical properties.

*1.1.1.2. Physical.* These include the coefficient of thermal expansion, which, if not adequately allowed for in design and assembly, can result in overstressing and distortion, particularly in the case of composite materials, in operation above or below the ambient temperature.

Thermal conductivity is of importance when heat transfer is being considered, e.g. jacketed vessels, condensers and other heat exchangers. Electrical conductivity must be considered when solvents of low flash point are being handled and static build-up must be avoided.

*1.1.1.3. Corrosion resistance.* This is obviously of particular importance for chemical plant and will be discussed in more detail later, but it is important that all types of corrosion are considered:

General wastage.
Localised corrosion, e.g. pitting and crevice corrosion.
Stress-induced corrosion, e.g. stress corrosion cracking, corrosion fatigue.

### 1.1.2. Appearance
This is not normally one of the main requirements for chemical plant, although it is important that surfaces can be easily cleaned, both internally and externally, and that external surfaces can be used for identification or warning purposes.

### 1.1.3. Fabrication

This is the ability of the material to be formed, welded or machined to the required shape. There is little point in selecting a material which meets all the other requirements, but which cannot be fabricated into the required plant item, except at greatly increased cost, or worse, where the processes involved adversely affect the very properties for which the material was chosen; for example, the adverse effect that welding can have in certain cases on the corrosion resistance of a material.

### 1.1.4. Compatibility with Existing Equipment

In a repair or maintenance situation there is often less opportunity for re-design than in construction of a new plant. Here delivery time and ease of fabrication and installation become of increasing importance. It is also necessary to take account of the remaining life of the plant or equipment so that repairs or replacements are not over-designed.

Other factors which may require consideration are:

1. The possibility of galvanic corrosion if dissimilar metals are coupled together.
2. The risks of thermite reactions with certain metals, and static build-up on non-conducting materials, in high fire risk areas.
3. The risk of zinc embrittlement of austenitic steels, again usually a problem only in high fire risk areas.

### 1.1.5. Maintainability

If an item is amenable to periodic inspection, to check for deterioration and, if necessary, to simple repair without undue effect on the operation of the plant, then the use of highly resistant, highly expensive materials is unlikely to be necessary, but such materials can often be justified for critical items on which the continued safe operation of the plant depends.

### 1.1.6. Specification Coverage

There was a tendency some years ago for companies such as ICI to provide their own materials specifications. However, more recently, there have been moves towards the use of national and international specifications and standards as far as possible with the advantage of relatively lower price, greater availability, increased rationalisation of materials used, and therefore reductions in stock requirements.

### 1.1.7. Availability of Design Data

This can be of particular importance for materials for duties at high or low temperatures and includes such factors as brittle fracture, creep and fatigue properties. The lack of proven design data can militate against the use of newer materials if the consequences of premature failure are serious enough.

## 1.2. Selection Considerations

### 1.2.1. Expected Total Life of Plant or Process

There is obviously no point in selecting a material for a life of 20 years if the plant on which it is to be used is to be knocked down next year; on the other hand, plant lives have a habit of extending rather than contracting from original estimates.

A greater problem may be changes in process life, particularly in the case of batch production, multi-product plant, as in the dyestuffs industry, where a capability to manufacture a wide and rapidly changing range of different products is often required.

### 1.2.2. Estimated Service Life of Material

This may have to be the same as the plant life, but often it will be possible and economic to replace an item periodically, provided that deterioration is predictable and can be monitored so as to avoid unexpected failure and to allow adequate planning of repair and replacement requirements.

### 1.2.3. Reliability

Here we must conider the safety and economic consequences of plant failure. Safety is obviously of paramount importance and any risks must be minimised. However, the economic requirements of reliability will depend on such things as the costs of repair or replacement in the event of an unexpected failure and also the cost of any lost production.

### 1.2.4. Availability and Delivery Time

This factor is more critical in a maintenance situation than during design and erection of a new plant, although for critical items the holding of long-delivery spares may be justified.

### 1.2.5. Need for Further Testing

It may be more economic to specify a material of known corrosion resistance than to spend money on extensive testing of cheaper materials. The delay to the project or the test programme itself may cost more than the saving between the two materials.

### 1.2.6. Materials Costs

Care must be taken when comparing costs to ensure that comparisons are valid, e.g. differences in heat transfer coefficients for different materials can affect the size and, therefore, cost of heat exchangers. Also variations in density and strength can have a significant effect on the weight and, hence, cost of materials required, for example for manufacture of a pressure vesel.

### 1.2.7. Fabrication/Installation Costs

The additional costs of casting, welding, forming, heat treatment, installation, supporting structures, etc., must be taken into account for valid comparisons to be made between materials for a particular duty.

### 1.2.8. Maintenance and Inspection Costs

These must be estimated, usually on the basis of past experience, for the total expected life of the plant, particularly in view of the present trend in increased labour costs compared to material costs.

### 1.2.9. Return on Investment Analysis

Having assembled all the different item costs, this is where the accountant comes into his own.

### 1.2.10. Comparison with other Corrosion Control Methods

A final decision should not be made until the possible use of cheaper materials of construction allied to the use of corrosion inhibitors, protective coatings or anodic/cathodic protection methods, with their own initial capital and on-going costs, has been considered.

The relative importance of the factors listed in Sections 1.1 and 1.2 will vary with the particular application. The dividing line between the two is very fine and some items could justifiably be moved from one to the other depending on specific circumstances.

## 2. CORROSION PERFORMANCE

Let us now look at the information required to predict the corrosion performance of a material.

## 2.1. Process Variables

Main constituents, identity and amount—this is usually fairly easy to determine.

Impurities (trace constituents), identity and amount—these can be more difficult to predict. Sampling and analysis may help, but if this is done on the laboratory scale some correlation with actual plant conditions is necessary.

Temperature—this can usually be measured, but care must be taken to measure at the most relevant position in the plant. It may be possible, and indeed necessary, to modify process operation, particularly in terms of temperature, to reduce corrosion to an acceptable level provided there is no adverse effect on other factors, e.g. composition or pH of process streams.

Degree of aeration—this can increase or decrease corrosion, particularly localised corrosion, depending on specific circumstances.

Velocity of agitation or flow rate—increases can cause removal of protective films and corrosion or erosion; on the other hand, decreases can result in stagnant conditions, leading to localised corrosion.

Pressure—this can affect process chemistry, e.g. dew point conditions and solubility limits, which may affect corrosion rates. It can also cause stresses in the material which may lead to mechanical failure (overload) or stress-induced corrosion.

Estimated range of each variable—this is particularly important and must cover not only normal operating conditions but also start-up, shut down and any foreseeable abnormal conditions. It may be necessary to choose a more resistant and usually more expensive material to cope with these conditions or to compromise in terms of suitability for normal operation to provide adequate resistance under abnormal conditions.

## 2.2. Type of Application

What is the function of the part or equipment?—Different materials are often chosen for different components exposed to the same environment, when properties other than corrosion resistance become important.

What effect will uniform corrosion have on serviceability? Is size change, appearance or corrosion a problem? e.g. build-up of corrosion product is normally unacceptable on heat transfer surfaces. Dissolution of metals by corrosion may result in contamination or discoloration of products. Localised corrosion is usually unacceptable since it is more difficult to detect and to monitor than general wastage. It can cause local weakening of a structure with little or no change in appearance. Stresses, both in-built during manufacture and operating stresses, can increase the risk of stress-induced corrosion.

Is the design compatible with the corrosion characteristics of the material? e.g. if the material is known to be prone to localised corrosion, have crevices (tube/tube plate gaps and backing ring welds) been designed out as far as possible? Also condensation traps, unlagged branches/dead legs/poor insulation should be avoided.

What is the desired service life? As already mentioned, this should not be under- or over-estimated.

What is the effect of cleaning? Often cleaning operations involve more arduous conditions than normal operation, particularly in terms of temperature, and care must be taken that these conditions are considered during the selection of materials, especially for multi-product plant where cleaning between campaigns is routine and can involve steaming and sterilisation operations.

What is the effect of time out of service? For example, between batches or campaigns, equipment may be exposed to more corrosive conditions than in normal operation because of air or moisture ingress and dew point conditions. If these periods are extensive then the need for protection by drying out, inert gas purging or corrosion inhibition should be considered.

What are the external conditions? Atmospheric pollution/contamination can present a corrosion risk. Corrosion by heat transfer media, e.g. steam, cooling water, brine must be considered. The effect of chloride-containing thermal insulation material on austenitic steel equipment should be assessed.

### 2.3. Experience

Has the material been used previously under *identical* conditions?
Is the equipment still in service?
Has it been inspected?
Has its performance been satisfactory?
Has the material been used in a similar situation?
What are the differences?
Are these critical in terms of likely corrosion behaviour?
What was the performance of the material?
Are there any plant corrosion test data?

Were the tests carried out in the area(s) of highest corrosion risk?
Did they include all proposed operating conditions?
Did they include the effect of heat transfer?
Have any laboratory tests been done?
Can these be related reliably to plant conditions?
What literature is available?

This can be a useful initial screen, but must normally be backed up with some form of corrosion testing under relevant conditions.

## 3. POTENTIAL PROBLEMS

Having looked at the factors we must consider in assessing the corrosion performance of a material, let us now consider the problems that can be met when using the more common materials for handling solvents. Table 1 distinguishes arbitrarily between dry solvents and those containing some water and gives some very general comments on the performance of materials, pointing out some particular hazards which should be avoided.

## 4. CONCLUSIONS

In general terms, the process of materials selection involves three important steps:

listing the requirements to be met by the material;
selecting and evaluating materials which are likely to be able to meet these requirements;
choosing from these the most economical material to serve the purpose.

Provided that full consideration is given to the first two steps then the final selection is relatively straightforward.

For selection of materials of construction for use in the manufacture and handling of solvents, consideration of corrosion performance is most important. However, the picture painted may be rather complex and a reminder that probably about 80% of chemical plant is still constructed in carbon steel is worth giving.

Nevertheless it can be seen that there are potential pitfalls even with the use of more exotic and expensive materials and it is not always

TABLE 1

The More Common Materials of Construction for Handling Solvents

| Material | For dry solvent | For solvent + water | Particular hazards |
|---|---|---|---|
| Mild steel | Satisfactory—most commonly used | Generally unsuitable if risk of hydrolysis to HCl or HNO$_3$ | Formation of corrosion product can lead to blockages, etc |
| Cast iron | | As mild steel | |
| Austenitic stainless steels | Satisfactory—used where cleanliness is important | Can suffer localised corrosion if HCl generated | Risk of stress corrosion cracking with aqueous chlorides above 60 °C |
| High nickel austenitic alloys, e.g. Incoloy, 825 | Satisfactory but not normally required | Improved resistance to pitting, good resistance to SCC | |
| Nickel & alloys — Nickel Monel 400 | as Incoloy 825 | Good—used for solvent distillation & recovery | Risk in solvent HCl/Cl$_2$ systems if solubility of corrosion product is high |
| Inconel 600 | Used for high temperature applications | | |
| Copper & alloys | Generally satisfactory —see hazards | Limited resistance under reducing conditions | Unsuitable for amines & solvents which can form acetylides, e.g. vinylidene chloride |
| Aluminium & alloys | Generally satisfactory —see hazards | Not resistant under acidic conditions | Unsuitable for certain completely anhydrous solvents, e.g. alcohols & some chlorinated hydrocarbons |
| Lead | Satisfactory | Resistant to dilute HCl but not HNO$_3$ | |
| Titanium | Satisfactory but not normally used | Satisfactory with cold dilute HCl but not normally used | Risk of stress corrosion cracking in anhydrous methanolic chlorides |
| Thermoplastics & thermo-setting resins, rubber | Not normally suitable for use with non-polar solvents. Can be used with acids contaminated with solvents below their saturation level | | |
| Enamel — Furane & phenolic resins — Fluorocarbons | Generally satisfactory within thermal limitations | | |

sufficient to spend more on the construction material to solve a corrosion problem. Many, if not all, of the other factors mentioned must be taken into account. The best way of ensuring that this is done, certainly in the case of a new plant, is to consider selection of materials at an early stage in the project so that materials and processes can be fitted to produce solvents and other chemicals as safely, reliably and economically as possible throughout the required life of the plant.

## REFERENCES

1. G. N. Kirby, *Chemical Engineering,* 1980, **87**(22), 86.
2. R. E. Moore, *Chemical Engineering,* 1979, **86**(14), 101; *ibid.,* 1979, **86**(16), 91.
3. L. Evans, 'Selecting Materials for Construction and Maintenance of Process Plant', presented at the Eurochem 80 Conference.
4. J. G. Hines, 'New Approach to Plant Corrosion Problems', ICI Internal Communication, July 1968.

PART III

# SOLVENT RECOVERY AND DISPOSAL: ECONOMIC AND ENVIRONMENTAL ASPECTS

# 13

## SOLVENT RECOVERY BY DISTILLATION*

C. PENNING

*Paktank Industriele Dienstverlening, Rotterdam, The Netherlands*

### ABSTRACT

*Recent increases in the price of chemicals and the growing concern about the disposal of chemical waste have made solvent recovery economically and environmentally more acceptable. Complete recycling is not always the best solution. Partially treated waste stream can offer an attractive economic alternative. Factors affecting the decision about solvent recovery are discussed, viz. solvent cost and future outlook, cost components of solvent recovery and disposal costs of used solvents. Examples are given to demonstrate cost analysis.*

## 1. INTRODUCTION

Recovery of solvents in economic terms means spending costs:

to return a waste solvent to its original or new application value;
to avoid environmental (disposal) costs.

Below, I shall deal with the changing economic factors influencing solvent recovery processes.

To begin with, I would like to outline the field of large-scale (more than 1000 tonnes/year volumes) industrial solvent recoveries and concentrate on the various types and combinations of distillations. This is not only because of my background in this field, the five years I spent in marketing distillation services (not only for recovery

* Presented at the 3rd Solvents Symposium for Industry in 1980.

151

purposes), to the Dutch and European industry, but also because distillation, although a well-known and conventional process, is still seen as a major way to purify polluted solvents.

The following sections will consider:

price changes of chemical solvents;
development of recovery costs;
factors influencing the earning power of solvent recycling;
the decision whether to develop solvent recycling internally or to process externally.

## 2. PRICE CHANGES OF CHEMICAL SOLVENTS

Since the first oil-supply crisis in 1973–74 the cost of most petroleum-based chemicals has increased sharply. No explanation of the reasons behind this is needed here. The only important factor for us at the moment is the clear perception of the increasing cost of solvents produced in the chemical industry by this event. The aftermath was a flattening, and even falling, of prices during the following four years of over-capacity in chemical bulk production. In 1978 the prices again came under the influence of the psychological effect of a possible oil-shortage and this raised chemical prices to a higher level once more.

Especially, the prices of chemicals directly derived from naphtha felt this effect. Under the influence of increasing naphtha feedstock cost the prices of aromatics rose to extreme levels (Fig. 1).

The price of benzene (Fig. 2) rose by more than 300% in 1974. It declined during 1975–77 and started to increase again during the year 1978, culminating in a price peak of US$620 per ton during December 1979. The same pattern is shown by the curves reflecting the prices of toluene and xylene.

Other solvents more downstream also had their increases, although many not so dramatic as their aromatic relatives due to the fact that distribution, packaging and production costs dim the effect of feedstock increases and because of a general tendency to overproduction in the chemical industry.

So far we can draw a few general conclusions:

solvents' costs will form a higher part of the total solvent application cost;

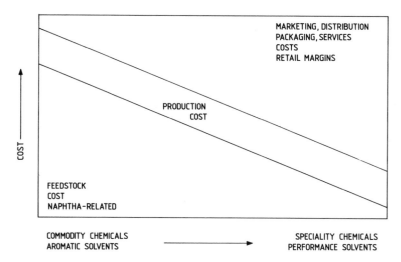

Fig. 1. Price changes of solvents.

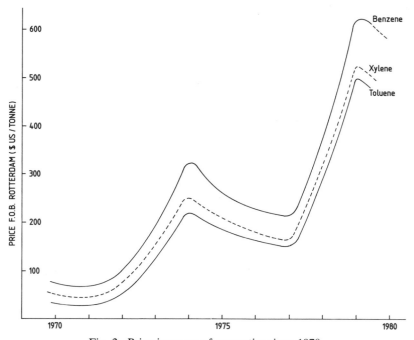

Fig. 2. Price increase of aromatics since 1970.

usually cheap solvents like aromatics will be replaced by performance solvents or non-oil based solvents (water);
more attention will be paid to the re-use of waste solvents.


## 3. RECOVERY COST AND ITS DEVELOPMENT

With the conclusion in hand that the price increase of solvents is the
'motor' effect for starting a solvent recovery project, it is interesting to
look at recovery costs, i.e. purifying a waste solvent stream by one or
more distillation steps, and the development of these costs during the
last few years. The relevant cost components for a distillation process
can be simplified as follows:

investment costs (fixed)
operational costs (variable/fixed)
   energy
   labour
   maintenance
   general

### 3.1. Investment Costs
Capital investment in a solvent recovery plant depends in part on the
situation of the main plant itself. For example, in a petrochemical
complex with a lot of additional equipment (steam/heating) available it
is less costly than building an isolated column in a metal working
factory.

In general, investment is a composition of the parts shown in Table
1.

The main cost components, mainly material and personnel costs,

TABLE 1

|                                                          | Fraction |
|----------------------------------------------------------|----------|
| Column including tube reboilers/condensers coolers/pumps | 0·3–0·5  |
| Heating system                                           | 0·2–0·3  |
| Control system                                           | 0·1–0·2  |
| Piping/tanks                                             | 0·1–0·2  |
| Engineering                                              | 0·1–0·15 |

TABLE 2

| Year | Capital investment cost index[a] |
|------|------|
| 1970 | 48·7 |
| 1975 | 100 |
| 1978 | 146·7 |
| 1979 | 162·8 |
| 1980 | 187·3 |

[a] Source: CIA, UK, 1980.

have risen sharply during the last 10 years resulting in an overall escalation of investment cost, shown in Table 2.

To give an idea of the total investment needed:

a column built in stainless steel and having 30 to 50 trays (enough to separate acetone/water, methanol/water), needs an investment of about Dfl* 0·8–1·2 million/tonne/h (working capacity 0·5–5 tonne/h), 1979.

For example: to process 3000 tonnes of solvents during 5000 working hours/year would involve an investment of Dfl 0·5–0·7 million.

## 3.2. Energy Costs

Energy, regrettably, is one of the main cost components of distillation and the amount of energy needed can fluctuate between 30% and 60% of total standard distillation costs.

The main energy consumer is the evaporation of the liquid mixture in order to get the necessary reflux for separation. The composition of the waste solvent as well as the required purity for the re-usable solvent determines the energy needed.

For example: a separation of 50% methanol from 50% water, separated in a 25 tray column, involves an evaporation/feed ratio of about 1. Separating 50% methanol from 50% water with an evaporation/feed ratio of about 2, in order to obtain a higher purity level, means about 90% more energy consumption.

The development of energy costs is shown in Fig. 3 by presenting the cost escalation of heating oil (heavy fuel oil) and the escalation of steam supply costs.

* Dfl—Dutch florin.

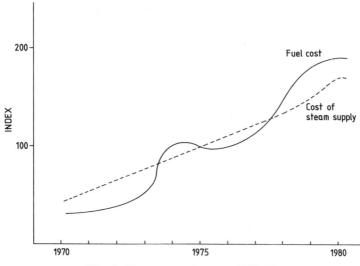

Fig. 3. Energy cost increase 1975–80.

### 3.3. Labour, Maintenance and General Costs

Labour requirements including maintenance also depend on the situation of a particular industry. A continuously operated distillation column requires the care of at least two people. If the recycling plant is built within the system of a petrochemical complex, no extra labour may be required. An isolated independently operated recycling plant needs the two extra people per shift and this can increase labour costs by up to 25–30% of the total distillation cost.

Taking all the components into account and trying to make a general estimate of the increase in recovery costs based on the distillation costs, it can be concluded from the cost split-up (Table 3) that the recovery costs have escalated by 70–80 per cent during 1975–80 (Table 4). This

TABLE 3
Distillation Costs Split-up (per cent)

| | |
|---|---|
| Energy | 30–50 |
| Labour | 10–20 |
| Maintenance/general | 10 |
| Capital | 30–50 |

TABLE 4
Distillation Cost Increases (per cent) 1975–80

| | |
|---|---|
| Based on replacement value | 70–80 |
| Based on historical cost value | 30–60 |

leads to the following intermediate conclusions:

(i) due to high energy requirements and the cost escalation of energy there is a strong incentive to develop energy conservation systems in distillation processes; the possibilities, however, are limited;

(ii) it may therefore be worth including the energy consumption during recovery in the total solvent application cost;

(iii) there is a tendency to develop other less energy consuming processes to recycle solvents.

### 3.4. Conclusions

Having discussed both the cost increase of solvents and the escalating recovery costs it is of interest to confront both developments and try to draw some general conclusions (Fig. 4).

As mentioned earlier, the cost increase of solvents over the period 1975–80 varied between 30% (speciality solvents, non-petroleum-based) and over 100% (aromatics, solvent naphthas), while the recovery costs, mainly by increasing energy costs, escalated by 70–80%.

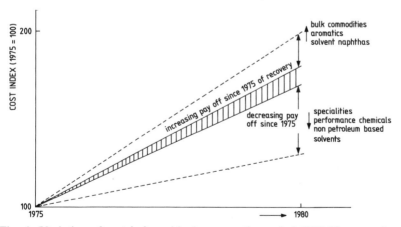

Fig. 4. Variation of cost index with time over the period 1975–80 comparing price increase of solvents against price increase of recovery costs.

It is obvious that it would be economical to recover more of the 'commodity' solvents—solvents with a high feedstock cost component—because of the increased earning power of the recovery process. On the other hand, and this is something that not many have realised, conventional ways of recovering speciality chemicals may have lost some of their attraction since 1975, due to a more than proportional increase of distillation costs against the increase in product costs.

## 4. SOME ASPECTS OF INFLUENCING EARNING POWER OF SOLVENT RECOVERY

### 4.1. Disposal Costs
So far we have not considered the influence of the disposal costs of waste solvents. Sometimes, if a nearby boiler house can be readily adapted, the waste can be used as an additional fuel.

In most cases, liquid waste has to be collected by a waste disposal company that has the equipment to dispose of it in a proper and legal way. However, this can only be done by spending money. In the Netherlands, prices for the disposal of liquid chemical waste range from Dfl 100–350/tonne.

### 4.2. New Use Value/Energy Costs
In some cases a waste solvent can be recovered and purified to a lower quality grade than that of the original solvent. When finding a new application for this 'low grade' solvent, the recovery can be rendered simpler and can save a lot of energy costs, while on the other hand the end-product can still approximate the value of the original solvent. This 'new use', or 'new application', can thus increase the earning power of a recovery. Two examples of these 'simple route' recoveries are (i) the recovery of methanol and the use of low-grade methanol in gasoline, and (ii) the use of water/solvents azeotropic mixtures for new solvent applications.

### 4.3. Uncertainty
The increasing political turbulence in the world has created more uncertainty about the availability of chemical feedstocks in the industrialised western world. Feedstock patterns and availability forecasts are changing rapidly and price fluctuations can have an extensive influence on applications of chemicals in general and solvents

in particular. It becomes very hard to predict what is going to happen in the next five or ten years, and if forecasts are being made, the range of probabilities (expressed, e.g. in standard deviations in a probability curve) will be broadened.

Companies which consider starting internal solvent recovery have to face this uncertainty. Since investing money in a certain system means converting the external uncertainty into internal company risks over the invested capital, the present level of uncertainty plays a major role in the decision to invest in a recovery system.

## 5. INTERNAL VERSUS EXTERNAL SOLVENT RECOVERY

Apart from technical and environmental arguments that play a role in such a decision, I would like to concentrate on the economic aspect of the decision by presenting and comparing two cases (Fig. 5).

*Case A*

*Waste:* 5000 tonnes/year
                DMF/water/polymer
Composition (%):   25/   73/     2
DMF-Value: Dfl 2400/tonne
Mode of operation: *Internal recycling by investing*

*Investment costs in year zero*[a]
Dfl (1·4 + 0·35) million = Dfl 1·75 million

*Operational costs per tonne intake*
  Energy             Dfl 120
  Labour             Dfl   50
  Maintenance
  General           __Dfl   25__
                     Dfl 195

Yearly return starting year 2:
((0·25 × Dfl 2400) − Dfl 195)
  × 5000 = Dfl 2 million/year

[a] 5·0 × 24 h/week − 12 h startup/completing during 40 weeks/year = 4300 h
Capacity=0·86 tonne/h
2 Column system

*Case B*

*Waste:* 5000 tonnes/year
                DMF/water/polymer
Composition (%):   25/   73/     2
DMF-Value: Dfl 2400/tonne
Mode of operation: *External recycling by contract processing*

*External recycling costs per tonne*
Including transport, handling, storage, interest on overquantity, insurance:

Dfl 350.

Yearly return starting year 1:
((0·25 × Dfl 2400) − Dfl 350)
  × 5000 = Dfl 1·25 million/year

Fig. 5. Internal versus external solvent recovery.

Bases: Planning horizon five years
Internal rate of interest
(cost of money)    8%

In Dfl million

### A.a

| | 1 | 2 | 3 | 4 | 5 | 6 | 7 year |
|---|---|---|---|---|---|---|---|
| Earnings | | +2 | +2 | +2 | +2 | | |
| Outlays | −1·75 | | | | | | |
| Discount factor | 0·93 | 0·86 | 0·79 | 0·74 | 0·68 | | |
| Discount cash flow | −1·63 | 1·72 | 1·58 | 1·48 | 1·36 | | |

Present value: 4·5

### B.a

| | 1 | 2 | 3 | 4 | 5 | 6 | year |
|---|---|---|---|---|---|---|---|
| Earnings | +1·25 | 1·25 | 1·25 | 1·25 | 1·25 | | |
| Outlays | | | | | | | |
| Discount factor | 0·93 | 0·86 | 0·79 | 0·74 | 0·68 | | |
| Discount cash flow | 1·12 | 1·08 | 0·99 | 0·93 | 0·85 | | |

Present value: 5·0

Fig. 6. Internal versus external solvent recovery—further details.

Case A represents a project proposal for internal recycling of a 5000 tonnes/year DMF (dimethylformamide)/water/polymer stream with a composition (%) of, respectively, 25/73/2, whereby DMF is recovered after treatment. (Further details are given in Fig. 6.) Case B represents the possibility for external recycling of the same stream, resulting also in DMF recovery. I would like to compare the two cases economically by taking into account only the real cash incomes and outlays during a certain year.

An investment is regarded as a net cash outlay in the year the investment is made, in contrast to normal accounting when a certain form of depreciation over the following years is made.

The other principle is that the 'cost of money' is taken into account by correcting all the cash flows in later years by a discount factor, determined by the interest rate the company in question is paying to have working capital at its disposal. Now let us take a further look at the different influencing parameters, viz. planning horizon, interest rate, operational costs and capital investment.

## 5.1. Planning Horizon
In the case of a short planning horizon (say less than five years), caused by a high level of external (and internal) uncertainty, external recycling would be more favourable (Fig. 6):

Present value + Dfl 4·5 million
     against + Dfl 5·0 million in the external alternative

Extending the planning horizon to over seven years would reverse the position.

## 5.2. Interest Rate
In the above example we used an interest rate of 8%, a percentage that was normal during the late 1960s and early 1970s. Now these interest rates have increased. The effect this has on investment decisions will be clear. In our example the present values become:

Case A: + 3·56 $\Big\}$ at an internal rate of interest of 14%
Case B: + 4·25

Thus, increasing interest rates lead to avoiding investments and making use of existing external equipment on a contract processing basis.

### 5.3. Operational Costs

When a company already has energy available in their internal system the question whether to process internally or externally hardly exists. In our case, it would keep the external route out of any consideration. The present value on a period of five years with a yearly income of +Dfl 2·6 million instead of +Dfl 2·0 million would be increased to 6·3. The same effect, only to a less extent, is applicable to labour costs.

### 5.4. Capital Investment

When recovery of a solvent mixture demands a very complicated way of processing, for example the use of more than two distillation steps to get a material with enough purity for use in the original plant again, the investment costs are relatively high and their influence on the present value of the investment route is heavier. Since the investment is to be made in year 1, the positive cash flows during the next years will need a longer period to pay back the invested money. In this case, one may consider turning to an outside specialist in the field of solvent recovery to develop an adequate process in their existing system.

# 14

# SOLVENT RECOVERY USING ACTIVATED CARBON

I. Metcalfe and C. S. H. Wilkins

*Sutcliffe Speakman Engineering Ltd, Leigh, Lancashire, UK*

## ABSTRACT

*Activated carbon, manufactured from coal, coconut or wood under controlled conditions, is characterised by the presence of molecular-sized pores and has a remarkable affinity for the adsorption of many organic molecules. The adsorption mechanism, including adsorption dynamics, and the factors affecting adsorption and desorption of solvent molecules are considered. The design of a modern solvent recovery plant using activated carbon is related to the economics of solvent recovery.*

## 1. WHAT IS ACTIVATED CARBON?

Activated carbon is a highly porous form of carbon or charcoal. Charcoal has been known to purify water for over 2000 years, and has been used to purify sugar solutions since the thirteenth century. In the late eighteenth century Scheel discovered the ability of charcoal to *adsorb* gases, but it was not until the period just before the First World War that it was used commercially in Germany to remove solvents from the drying air in the printing process. At about the same time in England, Sutcliffe Speakman pioneered the use of activated carbon for use in gas masks and later, in the 1930s, installed the first solvent recovery plant for recovering the solvents emitted during the rubber coating of textiles. Solvent recovery plants, using activated carbon are now in use throughout the world, and are indispensable in the publication gravure, magnetic tape, and many other industries on both economic and environmental grounds.

Activated carbon is produced by burning a suitable feedstock, such as petroleum fractions, coal, wood or coconut shell, in the absence of oxygen and then subjecting it to high pressure steam. Heating the carbon to 170 °C effectively dries the material, and subsequent heating to about 275 °C decomposes organic matter to carbon, driving off the non-carbon portion. The carbonised product is heated to between 750 and 950 °C by superheated steam, which passes through the carbon, expanding and extending the pore network.

The final product may be produced in either granular or pelleted form depending upon its final application.

## 2. ADSORPTION

The adsorption mechanism is predominantly caused by intermolecular (Van der Waals) forces which cause cohesion in solids and liquids. It is important to note that no chemical reaction takes place; indeed the process may be viewed as a condensation process. This is supported by the heat of adsorption being approximately of the same order of magnitude as the heat of condensation.

The potential theory of Polanyi (1914) describes the factors affecting adsorption and a more useful modified form is given by:

$$\text{Adsorptive capacity, } W = f\left(\frac{V_m}{T \log (C/C_i)}\right) \tag{1}$$

where $W$ is the g solvent per g adsorbent; $V_m$ is the liquid molar volume at boiling point; $T$ is the absolute temperature; $C$ is the concentration of saturated vapour; $C_i$ is the initial concentration of vapour.

This equation shows that adsorption is increased by:

Decreasing temperature
Increasing molar volume
Increasing vapour concentration

The adsorptive capacities of various types of carbons are normally presented as a plot of $W$(g/g) versus $C/C_i$ at constant temperature (isotherms). To compare carbons, the equivalent surface area may be used, but more usefully the weight per cent adsorbed of either benzene or carbon tetrachloride under standard test conditions is utilised.

By varying the production process and the raw material, carbons with differing pore-size distribution can be produced. The retentivity of solvents on carbon varies with the pore-size distribution as well as the molar volume of the solvent. This enables the designer to select the carbon which gives the optimum adsorption characteristics for a given solvent, bearing in mind cost, density and availability.

## 3. ADSORPTION DYNAMICS

Solvent-laden air (SLA) passes through the carbon bed with decreasing concentration of the solvent. The band in which adsorption is occurring is known as the mass transfer zone (MTZ). The concentration gradient set-up can be represented as a wave front, and as the successive layers of carbon reach saturation the MTZ gradually moves through the bed until solvent is detected in the effluent air. This 'breakthrough point' is used to determine the bed capacity.

The MTZ represents the minimum bed depth for a given system and may be determined by:

$$\text{MTZ length} = \frac{\text{Total bed length}}{t_s/(t_s - t_B) - x} \qquad (2)$$

where $t_s$ is the time to reach saturation; $t_B$ is the time to breakthrough; $x$ is the % saturation in MTZ. The maximum bed depth is determined by the amount of solvent required to be held by the bed.

## 4. FACTORS AFFECTING ADSORPTION

Adsorptive capacity of a carbon bed is dependent on the physical and chemical properties of the solvent and the carbon selected, but a number of factors must be considered:

(1) The MTZ increases in length with increasing superficial gas velocities. The accepted maximum velocity to avoid fluidisation of the carbon is 75 fpm.
(2) Adsorption increases with decreasing temperature.
(3) High relative humidity decreases adsorptive capacity and increases the MTZ.
(4) Capacity increases with increasing concentration or partial pressure of the solvent.

(5) Carbon, a non-polar material, preferentially adsorbs non-polar compounds.
(6) Multi-solvent mixes extend the MTZ and reduce adsorptive capacity.
(7) Low boiling solvents will 'breakthrough' first in mixed solvent systems.
(8) Adsorption increases with increasing pressure.

## 5. FACTORS AFFECTING DESORPTION

To reverse the adsorption mechanism the equilibrium described by eqn (1) must be reversed, i.e. by increasing the temperature and decreasing the solvent concentration in the gas by either decreasing pressure or by purging. Bearing in mind the need to heat the carbon and the containing vessel, a number of alternatives are available:

|  | *Percent desorbed* |
|---|---|
| Indirect heating at 100 °C | 15 |
| Reduced pressure at 20 °C | 25 |
| Gas recirculation at 130 °C | 45 |
| Direct steaming at 100 °C | 98 |

Steam is the most efficient heat carrier available, is in the gaseous form enabling solvents desorbed to be carried away, and is easily condensed to recover the solvent as a liquid.

To prevent excessive wetting of the carbon bed it is desirable to have a few degrees of superheat in the steam. Steaming is carried out in the opposite direction to adsorption to avoid diffusion of the solvent throughout the bed.

## 6. SYSTEM DESIGN

Figure 1 illustrates a typical modern solvent recovery plant. Solvent-laden air is passed through filters (to remove dust) and either heaters or coolers to bring the air to approximately 38 °C. Where appropriate, heat recovery exchangers are used to reduce energy requirements. The air then enters the inlet air manifold via the fan; automatic, pneumatically controlled valves allow the air to pass through the

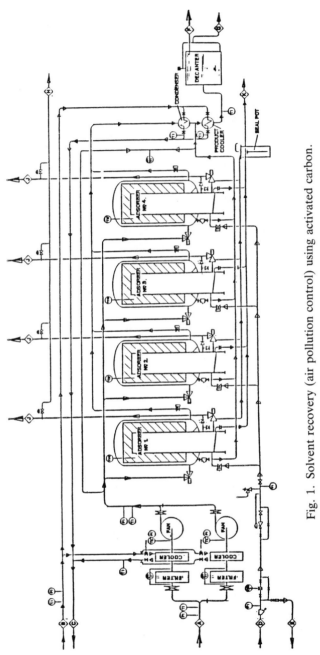

Fig. 1. Solvent recovery (air pollution control) using activated carbon.

adsorber on stream. The adsorber contains a bed of activated carbon between 18 in and 36 in deep arranged in either a horizontal or vertical annular bed. As the air passes up through the carbon the solvent is adsorbed, allowing clean air to pass to the atmosphere via the exhaust valve and chimney. A gas analyser samples the exhaust gas and when solvent is detected, at a predetermined level, the carbon bed has reached maximum capacity. The solvent-laden air is then diverted to a second adsorber and, by switching the appropriate valves, atmospheric pressure steam is fed to the saturated adsorber. The steam flows countercurrent to the SLA. The first few minutes of steaming are used to heat the bed and adsorber, the condensate being fed to the seal tank which also acts as a pressure relief system. Once steam has broken through the carbon bed it is passed, together with solvent vapour, to the condenser and cooler heat exchangers. The condensate is collected in the decanter where water-immiscible solvents are separated off for re-use. Water-miscible solvents are sent to a distillation plant for water removal. The adsorbers cycle automatically and the sequence is controlled and monitored by a central microprocessor.

Heat recovery from the condensing steam/vapour mixture is commonly used to preheat boiler feed water and provide process hot water.

The quantity of steam required to recover the solvents varies with the solvent and the concentration of the SLA. However, most modern plants use between 2 and 3 lb of steam per lb of solvent recovered.

## 7. LEL CONTROL

Most insurance companies will allow solvent recovery plants to operate up to 25% of the 'Lower Explosive Limit' without controls, but above that level monitoring and alarm devices are required. Solvent recovery plants use the least energy when the incoming air is at a constant flowrate and high concentration. By recycling process air until the concentration level is able to be maintained at as high a level as possible the heating requirement for the air is also reduced. By the combination of a continuously sampling gas analyser and automatic damper control complete safety can be maintained. The dampers are designed with a 'cut off' so that a minimum flow of air is always allowed to flow.

## 8. ECONOMICS

The justification for installing a solvent recovery plant utilising activated carbon is a reduction of operating costs for the process concerned.

| | |
|---|---|
| Air flow | 30 000 m³/h |
| Solvent load | 300 kg/h |
| Solvent cost | 40 p/kg |
| Steam cost | 1 p/kg |
| Water cost | 0·03 p/kg |
| Electrical cost | 3 p/kWh |
| Capital cost | £200 000 |
| Installation | £150 000 |
| | £350 000 |

Operating cost, for 4000 h/year

| | |
|---|---|
| Steam | £30 000 |
| Power | £7 680 |
| Water | £22 500 |
| Maintenance | £45 000 |
| | £105 180 |
| Recovered solvent | £465 600 |
| Depreciation | £35 000 |
| Interest on capital | £56 963 |
| Yearly operating costs | £197 143 |
| Net operating profit | £268 457 |
| Simple payback | 16 months |

## 9. CONCLUSIONS

The use of activated carbon for the removal of hydrocarbon vapours from an exhaust air stream is

(a)  a cost effective method;
(b)  simple and reliable;
(c)  well proven with plants opeating for over 30 years.

Normally, installation of an activated carbon adsorption system requires no modification to existing equipment since all that is required to be done is to duct the exhaust from the system to the inlet of the recovery plant.

To maximise the cost effectiveness of the installation of a recovery system the following areas have to be investigated.

1. Use of high concentrations of hydrocarbon vapours in air, i.e. up to 40% of the lower explosive limit.
2. Running at steady-state conditions, i.e. minimise machine downtime.
3. Use of water-immiscible solvents wherever possible which removes the need for extra distillation equipment to recover the solvents from water.

## REFERENCES

1. P. N. Cheremisinoff and F. Ellerbusch, *Carbon Adsorption Handbook*, Ann Arbor Science, Michigan, 1978.
2. T. K. Ross and D. C. Freshwater, *Chemical Engineering Data Book*, Leonard Hill (Books) Ltd, London, 1962.
3. F. R. Houghton and J. Wildman, *Chem. Proc. Eng.*, 1971, **52**(5), 61.
4. R. D. Hill, *Processing*, 1979, **25**(11), 86.

# 15

# HEAT RECOVERY FROM SOLVENT INCINERATION

W. G. Murcar

*Kaldair Ltd, Feltham, Middlesex, UK*

and

J. C. Boden

*BP Research Centre, Sunbury-on-Thames, Middlesex, UK*

## ABSTRACT

*As a method of reducing the operating costs of a hot air drying system, utilisation of the combustion energy of the evaporated solvent can be a profitable alternative to solvent recovery. This is particularly true if the recovered solvent would be unsuitable for immediate reuse.*

*Unless there is a use for surplus energy, the incinerator should be designed to minimise support fuel usage. A 'vortex combustor' will be described to illustrate means by which this may be achieved. Further support fuel reduction can result from low temperature catalytic combustion, provided no poisons are present, or from recuperation using the flue gases to preheat the incinerator inlet stream. Research with a unique recuperative furnace design will be outlined which shows that the extent of preheat which is feasible depends on the minimum oxidation temperature of the solvent or support fuel, which in turn depends on its composition.*

*Economic benefits are generally greatest with continuous coating processes, but where incineration is a requirement to reduce air pollution or odour the same principles may be applied to any solvent-laden stream to limit the fuel needed for incineration.*

## 1. INTRODUCTION—COST SAVING TECHNIQUES FOR SOLVENT DRYING PLANT

The fuel consumption of a typical solvent drying process can be reduced, sometimes almost to zero, by recovering heat from the combustion of the solvent-laden air. The technology required to achieve this is the subject of this paper.

This technique is one of a number of possible methods of reducing the operating costs of a hot air solvent drying system:

1. Reduce the air flow and monitor the solvent level.
2. Recirculate part of the exhaust.
3. Preheat the air by heat exchange.
4. Recover the solvent if reusable.
5. Burn the solvent-laden exhaust and recover heat.

Method 1, reduction of the air flow, will normally be the first step, provided that the drying process does not suffer. The solvent level in the exhaust is monitored as a fraction of the lower explosive limit (LEL).

Method 2, recirculation of part of the exhaust stream, is an alternative first step if the flow of air through the oven has to be so high that the solvent content of the exhaust after a single pass is well below the maximum safe level. This level is often set at 25% of the LEL, but may sometimes be raised, perhaps to 50%, if continuous monitoring is installed. Care has to be taken that recirculation does not lead to an undesirable build-up of any component. Recirculation will lower the oxygen level, partly compensating for the increase in the apparent % LEL of the solvent.

Method 3, partial preheat of the inlet air by heat exchange with the hot exhaust, is the first to involve significant additional hardware, and its economics will normally be compared with those of methods 4 and 5 which generally yield greater gains.

Methods 4 and 5, then, offer the greatest potential for savings, particularly when the solvent content of the exhaust is relatively high, and are also the only two methods to have an environmental advantage; they reduce the amount of solvent vented to the atmosphere.

Method 4, solvent recovery, will normally only be considered when the solvent is not permanently contaminated and would be economically reusable after any purification or separation costs have been taken into account.

Method 5, combustion of the solvent-laden exhaust, is discussed in this paper. A typical oven exhaust containing solvent at 25% of the LEL has a calorific value sufficient to provide a temperature rise of over 300 °C. Thus, there is in principle more than enough energy available from combustion of the solvent to provide all the heat needed by the ovens, and this will be demonstrated later.

In many industries, economic gains can also be provided by incineration of solvents in liquid form. In general, however, the technology needed to burn liquid solvents alone is simpler than that required to incinerate dilute solvent–air mixtures and will not be considered further here.

## 2. COMBUSTION OF SOLVENT-LADEN AIR

Combustion of solvent–air mixtures utilises the principles of incineration technology. In this paper, the ways in which the incinerator can be integrated with the drying ovens and with other on-site heating requirements in order to provide optimum heat recovery will be summarised. The general principles on which incinerators are designed will be described, showing how the combustion of very dilute fuel gas–air mixtures requires a specialised approach. This will be exemplified by a vortex-type low calorific value gas combustor, by an account of research into the highly recuperative combustion of dilute gas streams, and by a summary of the features of catalytic incinerators. In conclusion, an example of a typical application of a vortex-type combustor will be given.

Figure 1 is a block diagram showing most of the features that might be included in a solvent–air incineration scheme. In any given situation the optimum system is likely to include only some of these. Several points are worth noting. First, unless heat recovery for other on-site requirements is desired, the support fuel needed for the incinerator should be minimised. This is because a 25% LEL stream alone contains sufficient energy for the oven heating. This point will be covered more fully in the description of individual incinerators, but it may be noted that a recuperator is one means of reducing support fuel. Secondly, and for similar reasons, it may be necessary to incinerate only part of the oven exhaust stream to provide all the heat needed,

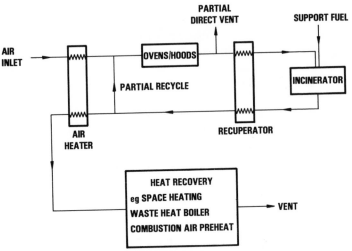

Fig. 1. Solvent–air incineration scheme showing all features which may be included.

the remainder being vented directly, environmental requirements permitting. Thirdly, the partial recycle and the air heater are normally alternatives. Providing combustion products can be accepted in the oven and do not affect the drying process, then recycle is cheaper and more efficient.

## 3. PRINCIPLES OF INCINERATOR DESIGN

Incineration requires heat provided either by the effluent or by the use of a prime fuel. Thus, with present and anticipated energy costs, it is logical to utilise the energy released as efficiently and as fully as possible. It will therefore be worthwhile considering the basic requirements of incinerator design.

Many people refer to the 'three T's' of incineration—time, temperature and turbulence. However, sufficient oxygen must also be present to give combustion/oxidation rather than pyrolysis, and thus we have the following four considerations: time, temperature, reactant concentration and mixing. Whilst these cannot be segregated into level of importance as they all play significant parts and are interrelated, they can be considered individually.

*Time:* The rate at which a reaction proceeds is critical to deciding the residence time required. The rate of reaction depends on the concentration of the reacting components, but for combustion reactions the rate of reaction is also temperature-dependent to a high degree.

*Temperature:* Temperature has a strong effect not only on the rate of reaction but also on the equilibrium position of the reaction. There is an additional major consideration: the higher the operating temperature, the more support fuel will be required. This is because the enthalpy of the incinerator exhaust gases will then be higher.

Therefore careful selection of the operating temperature with respect to both thermal efficiency and incineration efficiency is required. This balancing is necessary to guarantee optimisation of capital and running costs.

*Reactant concentration:* In most cases of incineration, the reactants are the effluent and oxygen (from air), and the reaction is a thermal oxidation. Then the quantity of oxygen available is critical. The equilibrium constant determines the final product concentration depending on the conditions acting upon the reaction: increasing the excess air factor, and thus the oxygen level, moves the equilibrium nearer completion, reducing the level of effluent in the exhaust gases. However, for solvent–air systems, providing the mixing is adequate, in general the oxygen is at such a high level that the operating temperature can be minimised.

*Mixing:* On the assumption that the time, temperature and reactant concentrations are satisfactory, it is necessary that mixing be good at all stages. Mixing can be induced in many ways, swirl being a well-proven method. However, mixing of gas streams, as opposed to gas–atomised-liquid mixing, can require considerable energy unless engineered correctly.

The level of mixing required can also be affected by the other parameters. At high temperatures, mixing is aided by rates of diffusion, and thus mechanical mixing is not so critical. Long residence times, or high oxygen levels, also reduce the mixing levels needed. However, good incinerators are not big, fuel-expensive, high oxygen level systems; on the contrary they should be as small and fuel-efficient as possible. This will ensure economy in capital and running costs.

In summary, flexible thinking in terms of mechanical design, plus correct balancing of time, temperature, reactant concentration and mixing, will result in simple, efficient operation.

Bearing all these aspects in mind, the specialised requirements of many solvent users and the equipment available to carry them out can now be considered.

## 4. THE VORTEX COMBUSTOR

Since in most drying oven exhaust streams the solvent to be burnt is already mixed with a large excess of air, the combustion equipment to be used should ideally be different from a standard incinerator. What is needed is a combustor specifically designed to burn dilute fuel–air mixtures. The separate support fuel burner used in conventional incinerators may even be eliminated altogether, extra gaseous fuel being added to the solvent stream if required.

A combustor designed for dilute fuel–air streams is shown in Fig. 2. It will be referred to as a vortex combustor—the design is based on work done at University College, Cardiff.[1] It illustrates well the techniques that can be used to stabilise the flame. First, the flow is strongly swirled in the primary chamber to produce a recirculation vortex in the second chamber, within which the flame stabilises. Secondly, the combustion zone is surrounded by high temperature refractory, or metal walls, which radiate heat back to the flame zone. The support fuel may be added to the incoming mixture, and, as a protection against extinction, there is a small continuously burning pilot near the centre of the vortex. The pilot can also be uprated to become a support fuel burner when this would be advantageous. Figure 3 shows a large-scale vortex combustor system designed to burn 1000 scfm* (1700 Nm³/h) of a gas stream with a total calorific value as low as 40–50 Btu/ft³ (1·5–1·9 MJ/Nm³). This is a demonstration/test unit used to optimise the design for different applications, and is built in sections so that the geometry can be altered as necessary.

## 5. RECUPERATIVE COMBUSTION

Since recuperation—preheating the incinerator feed by heat transfer from the flue gases—reduces the support fuel required to bring the stream to the combustion temperature, the question arises as to

---

* Standard cubic feet per minute.

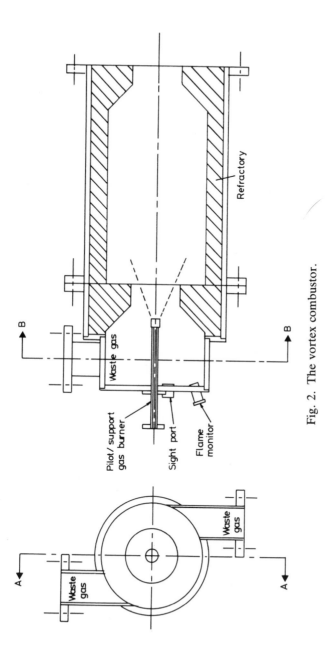

Fig. 2. The vortex combustor.

Fig. 3. Vortex combustor demonstration unit.

whether, with a typical oven exhaust stream, it is possible to incorporate such a high degree of recuperation that the support fuel can be eliminated altogether. This is clearly a worthwhile target, if it is technically possible and can be achieved for an acceptable capital cost. Oven fuel costs might even, during normal operation, be reduced to zero, all the heat required being provided by the solvent.

It has been shown that for methane the lean flammability limit can indeed be lowered indefinitely as the preheat temperature is increased; that is, any stream containing methane and sufficient air, no matter how dilute, can in principle be fully oxidised without support fuel by employing enough recuperation. Can this be extended to other common dilute fuel streams, and in particular to dilute solvent–air mixtures?

In an attempt to answer this question, a pilot-scale recuperative combustor was built. This was based on a design used, in the laboratory, by Professor Weinberg[2] of Imperial College to investigate the same problem of the combustion of very dilute fuel streams. In order to reduce heat losses and assembly costs and to increase operational flexibility, the burner and recuperative heat exchanger are integrated as a single unit. The system is also unusual in that the heat exchanger takes the form of a double spiral, or swiss roll, again to reduce heat losses, and also to provide a long path for countercurrent heat transfer with a moderate pressure drop. This type of heat exchanger is manufactured on special machinery by several firms, normally for lower temperature applications.

Figure 4 shows a cross-section of the combustor, together with a diagram indicating mass and heat flows. Although combustion ideally takes place mainly in the centre of the spiral, it can, if less than the full preheat is required—for example, if the stream becomes richer—move back into the inlet spiral, since the fuel and air are fully premixed before the inlet to the heat exchanger.

Figure 5 shows the scale of the experimental unit, which can burn 400 scfm (680 Nm$^3$/h) of 25% LEL methane–air mixture, releasing 88 kW.

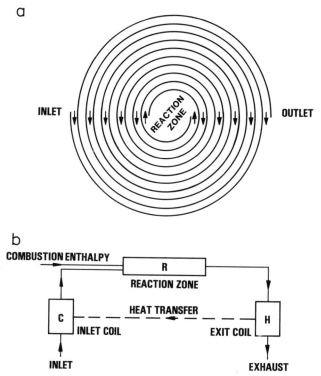

Fig. 4. (a) Schematic representation of the gas flow in a spiral burner; (b) principles of recuperative combustion.

Fig. 5. Prototype spiral burner.

The test results indicated that natural gas and propane, typical support fuels, could indeed be burnt unaided at least down to the 25% LEL level, corresponding to about 13 Btu/ft$^3$ (0·5 MJ/Nm$^3$), as could, for example, iso-octane. Figure 6 shows the temperature and the percentage heat release, calculated from the composition of gas samples extracted from the burner, as a function of distance along the inlet spiral, for propane at 21% LEL. Clearly combustion took place almost exclusively in the central chamber—between points 9 and 10—and the spiral heat exchanger was used to its maximum extent. The apparent preheat temperature was 650 °C.

Typical solvents, *n*-heptane, methyl ethyl ketone (MEK), isopropanol and white spirits, have also been tested, and although these start to oxidise at a lower temperature than the fuels listed above,

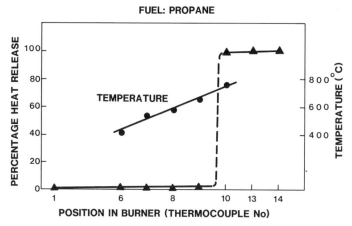

Fig. 6. Variation of temperature and heat release through spiral burner. Fuel—propane.

addition of quite a small proportion of one of these fuels again results in complete combustion of a very dilute stream. Figure 7 is a similar graph in which 25% of isooctane has been added to *n*-heptane to give a total calorific value of 15 Btu/ft$^3$ (0·6 MJ/Nm$^3$).

It is interesting to note that the tendency to oxidise at a lower temperature in these very dilute, relatively long residence time systems

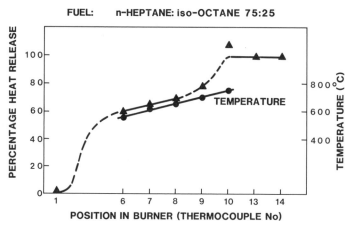

Fig. 7. Variation of temperature and heat release through spiral burner. Fuel—*n*-heptane : isooctane (75 : 25).

correlates neither with the normal auto-ignition temperature of the fuel, which is the temperature at which a stoichiometric mixture will initiate a flame, nor, for example, with its octane number, and so is difficult to predict from the literature. Therefore, since there is generally no benefit in attempting to preheat a premixed stream to above the temperature at which a significant oxidation begins, experiments of this type are vital if the maximum useful degree of recuperation is to be employed.

Although the spiral burner itself is still in the development stage, these results can be applied to conventional recuperative systems, using heat exchangers coupled with burners, such as the vortex combustor, enabling them to employ the highest feasible preheat temperatures and thus to use the minimum of support fuel.

## 6. CATALYTIC COMBUSTION

The techniques that have been described rely on normal gas-phase combustion. By carrying out the combustion on the surface of a suitable catalyst, it is often possible to burn dilute solvent–air mixtures at much lower temperatures. Thus the preheat required is reduced, and this is another way of saving on support fuel. The most usual

Fig. 8. Catalytic combustion test rig.

catalysts are various transition metal oxides or the platinum group metals, and, for the latter, combustion temperatures for typical solvents range from around 250 °C for simple aromatic hydrocarbons to around 350–450 °C for MEK or ethyl acetate. By adding some recuperation it is sometimes again possible to eliminate support fuel completely during normal operation.

Situations where catalysts can be used are those where it is known that no materials are present in the gas stream which could poison or coat the catalyst surface, and where it is not possible for the catalyst to be grossly overheated. Figure 8 shows a test rig we use to check the compatibility of specific fuel streams with catalysts and to measure catalytic ignition and combustion temperatures and required residence times for new gas mixtures.

## 7. SUMMARY AND APPLICATION EXAMPLE

The concepts outlined in this paper can be summarised with reference to the generalised block diagram (Fig. 1):

An incinerator designed for dilute fuel–air streams will minimise support fuel usage and will probably not require extra combustion air for a support fuel burner. Fuel usage can be reduced further by employing recuperation to the maximum extent feasible, or in suitable cases by using a catalytic incinerator. Any excess heat above that transferred back, either by air preheat or exhaust recycle, to the drying ovens themselves might be recovered and used elsewhere on the site.

Finally, the application of these principles is illustrated by taking the example of a resin coating plant in which 21 000 kg/h of air passes through drying ovens at an inlet temperature of 200 °C. These operate for an average of 6000 hours per year, and the fuel cost for direct gas-fired heating at 31 p/therm was £30 000. In addition, space heating, operating for 50% of the year, cost £11 250 in natural gas. This, together with a desired 40 °C increase in oven inlet temperature, gives a total gas bill of £48 000 per annum. The oven exhaust contains 0·28% (by volume) of methyl ethyl ketone, equivalent to 16% of the LEL. Because of uncertainty in the composition of future coating resins, a catalytic incinerator was not favoured. Therefore a vortex combustor with a recuperator was selected, and the oven air was heated by heat exchange rather than partial recycle. Finally, fresh air for space heating was warmed by heat exchange with the exhaust stream. The total duty

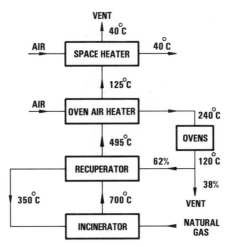

Fig. 9. Coating plant incineration system.

could be provided by incinerating only 62% of the oven exhaust, and the scheme is shown in block form in Fig. 9.

The fuel cost for the new system is estimated as £18 000 per annum, a saving of £30 000, or 63%, on the original cost.

Should it be feasible to run the ovens with a solvent concentration at the exit of 28% of the LEL, then, apart from a small pilot burner, no fuel would be required after start up. Then £45 000, or over 90% of the original fuel cost, could be saved.

## ACKNOWLEDGEMENT

Permission to publish this paper has been given by The British Petroleum Company plc.

## REFERENCES

1. T. C. Claypole and N. Syred, *J. Inst. Energy*, 1982, **55**(3), 14.
2. S. A. Lloyd and F. J. Weinberg, *Nature*, 1974, **251**, 47.

# 16

## SOLVENT DISPOSAL*

L. E. BAKER

*Re-Chem International Ltd, Southampton, UK*

### ABSTRACT

*It is accepted that, wherever possible, waste solvents should be recovered, or used as some sort of fuel. However, if this is not possible, then the only alternative method generally accepted within Great Britain is incineration. Landfill methods are not approved for inflammable materials and sea disposal is usually limited to predominantly aqueous wastes. More exotic methods, such as catalytic oxidation, find a place in emission abatement circles, but do not offer a realistic alternative to incineration for the disposal of conventional waste products.*

*Incineration is considered under three principal headings, the economic implications of each being emphasised. The three areas of discussion relate particularly to the emission standards required of process plants of this type, the arduous duty demanded of plant handling a wide variety of materials, and the tight administrative and quality control checks which must be instituted to comply with best practice and regulations.*

## 1. INTRODUCTION

The title, containing the two words 'Solvent Disposal', seems simple enough, but in a practical sense at least, both are capable of a number of interpretations.

* Presented at the 3rd Solvents Symposium for Industry in 1980 and revised in 1983.

As regards 'Disposal', I shall deal mainly with incineration, the most widely used method of solvent disposal, and review briefly other disposal techniques, such as landfill, sea disposal and catalytic oxidation. I shall also dwell, where possible, on economic aspects of solvent disposal.

If I were to interpret 'Solvents' in its most simplistic sense, that is to say mixtures of organic solvents which for some reason could not be recovered, but which were substantially free of solid contaminants, water, etc., and which were applied in drums or even bulk loads, then there really is only one significant disposal route available—incineration! However, bearing in mind a more broadly based review of available methods, I shall adopt a wider definition of 'solvent', to include solvent-containing wastes, residues from solvent processes and solvent vapours.

## 2. INCINERATION

Incineration is, and is likely to remain for some time, the principal disposal route for non-recoverable, flammable and toxic solvent waste. Indeed, that applies for many difficult organic wastes that are toxic, malodorous or of persistent nature, whether or not they are solvent or solvent-based.

The commercial basis for incineration varies from units built and operated by major waste-producing companies for their own exclusive use, to multipurpose contracting operations taking business from many sources. These latter operations vary from private-sector financed and operated incinerators competing in an open market against other disposal options (e.g. UK) to publicly funded operations working in virtually monopolistic conditions (e.g. some European countries).

Many factors influence the type of incinerator which may be needed, and the cost of disposal of any particular waste. The most important of these are:

—The physical form of the waste (i.e. liquid, sludge, solid).
—The calorific value.
—Its chemical composition with particular reference to contaminants, such as halogens, sulphur, nitrogen and toxic metals.
—The quantity and nature of any ash which may be produced.

Not surprisingly, combustion of straightforward waste solvents is technically about as easy a problem as one is likely to come across in incineration. Introduction to the incinerator via burners, with good atomisation and burning characteristics, represents long-established technology. However, except possibly in captive or in house markets, such straightforward applications are rarely found and more complex waste mixtures must be catered for. Solvent sludges and slurries pose problems of storage, pumping and atomisation, and solid residues, such as thermosetting still-bottoms or other solid wastes, require a whole range of fresh approaches to pre-treatment, sorting, and feeding into the incinerator.

In a chapter that is meant to dwell on the economic aspects of disposal practices, it is probably most convenient to break down the discussion into three areas:

  (a)  emission standards and the need for gas cleaning plant of various types;
  (b)  the arduous duty demanded of plant handling a wide variety of materials;
  (c)  the tight administrative and quality control checks which must be instituted.

### 2.1. Emission Standards and Gas Cleaning

Incineration of waste is a registrable process under the Alkali Act and it thus follows that the operation is under the surveillance of the HM Industrial Air Pollution Inspectorate. This Inspectorate sets standards of emissions in respect of particulate, gases, etc., and is entitled to pursue the concept of 'best practicable means' with the plant, facilities and procedures available.

Unless an incineration activity is confined to relatively clean liquid wastes, and wastes containing no components able to produce more than nominal amounts of acidic combustion products, then gas cleaning plant is going to be needed. This will add substantially to the running cost by way of much enhanced power and raw material requirements, additional labour, supervision, and increased maintenance work, etc. Typically, a gas cleaning plant may require several thousand gallons of pure water per hour to make up for evaporative loss, and for that portion of contaminated gas washer liquors which needs to be withdrawn to maintain dissolved and suspended solids equilibrium. The disposal of these liquors is another cost area. Neutralisation of acidic liquors produced in a gas washer will require supplies of lime, caustic soda, etc., at prices in the range of £30 to £80 per tonne.

Indeed, with highly chlorinated liquid wastes, incineration in sea-going ships, where gas cleaning provisions can be entirely dispensed with, can sometimes become a very attractive alternative compared to a landbased incinerator, with all its additional controls and consequent costs. There are a number of operators of sea-going incinerators basing their business generally on the incineration of highly chlorinated wastes. Activities in European waters are on the decline in line with the officially unfavourable view of this practice adopted by a number of governments.

Nevertheless, it is in the area of highly chlorinated wastes that the inherently high costs of running a ship-bound operation can be more than offset by savings in the avoidance of gas cleaning requirements.

Even incinerators equipped with gas cleaning plant, which is effective in removing acidic gases, particulate, etc., to within the prescribed limits, can still face problems. Dense steam plumes arising from the massive amounts of water injected to cool the combustion gases can still be visually obtrusive and can be stabilised by the residual low levels of particulate. To tackle these areas requires still more efficient particulate removal and, possibly, even some degree of reheating of the gas stream prior to final stack discharge. Solutions, whilst possible, are energy and capital intensive and can only further increase disposal charges for plants incorporating such steps.

For example, consider the possibility of adding a bag filter to an existing operation, with the intention of substantially reducing the particulate emission levels. A considerable number of horsepower would be required to pull the volumes of air through such a resistant installation, and the disposal of dust collected would be an additional expense. Bag life is difficult to predict, but even with a two-year replacement cycle, the cost of the bags and of the labour in doing the change is likely to add several pounds to each tonne of material incinerated.

Finally in this section, one must take account of the relationship between high combustion temperatures and the quality of the emission. Combustion temperatures in excess of 1000 °C, and with a residence time at that temperature of at least several seconds, are required if the more difficult and persistent organic species are to be totally broken down. Extensive and expensive testing has been undertaken to prove that this is the case, but a continuing financial burden is felt in maintaining supplies of reasonable CV (calorific value) liquids to ensure these high temperatures. Solvent wastes, such as may be usable as fuels in the incinerators, cannot normally be obtained at better than free-of-charge, the revenue from the incineration operation as a whole being obtained from the disposal of solids and more difficult species.

## 2.2. Arduous Duty

In-house incinerators may have an advantage over contract units in this area, but in general the tremendous range of wastes handled (in both physical and chemical terms) makes enormous demands on the fabric of the plant installation. It is the contractor's lot to handle the business that comes to him (and which he is capable of doing effectively). Wear and tear on pumps usually through corrosion or blockage, corrosion of pipes and tanks, excessive wear and tear on the refractories, combined possibly with slagging problems, short life of instrumentation, arduous duty and consequent short life of fork trucks, etc., seem to be the feature of all multipurpose incinerator operations in the UK, Europe and North America.

Major refractory replacement and repairs programmes are more or less an *annual* feature of many big operations, and minor repairs are almost always building up for programming into intermediate shut downs.

With reference to Re-Chem, our records show that for each tonne of waste incinerated during 1982, the average repair and maintenance costs were about £9. This sort of figure has been consistently recorded over recent years, and is also typical of figures which we have obtained for other incinerators both in the UK and Europe.

## 2.3. Administration and Control

The lot of a contract operator is not a happy one, and it is certainly not a profession I would recommend to anyone wanting an easy, orderly and predictable life. We appear to be affected by a bewildering array of regulations and inspecting officers, each of whom has his own sets of forms, registers and records to be maintained. More than that, however, the contract operator has the onerous responsibility of attempting to sustain his credibility in an area where reputations are not all that they might be. The tendency, hopefully now on the decline, to assume that a disposal contractor is something of a rogue, is one that the responsible contractor endeavours to live down by expending considerable effort and care in maintaining a controlled operation. The statutory procedures represent only the minimum that needs to be done, and in practice the responsible operator must go beyond that if he is really to establish a reputation of the highest order.

What does this mean in 'economic' terms? Certainly, we need to employ a number of staff who are not directly concerned with the operation, but who are employed in cross-checking and co-ordinating all the information necessary to maintain a secure operation. This involves laboratory personnel checking by analysis the wastes received at the plants, office-based clerks processing statutory notifications and cross-checks, and plant-based production planners and process controllers who sort through, categorise, store, and maintain 'life history' records of each batch received. It is clear that such personnel will be expensive, and will contribute significantly to the overheads of the operation. They can be skimped on, but only at risk of the standard of the operation falling, and mistakes being made.

## 3. LANDFILL

As I hinted in the introduction, landfill of solvent waste should not be used. Generally, solvents will either be flammable, thus posing obvious problems to landfill sites, or halogenated, thus posing particular groundwater pollution threats.

Needless to say, solvents will be landfilled in conjunction with other wastes, e.g. with paint sludges, plastic residues, resins, etc., but not normally with much 'free' solvent around. Such materials could, of course, be incinerated, and indeed very often are, but it would be foolish to pretend that where landfill is allowed to be practised, then it will not represent a significantly cheaper option.

Nevertheless, landfill activities have experienced significant increases in control, and improvement in standards. Site licensing and the various notification procedures required have introduced some administrative responsibilities, and landfill management practices have usually greatly improved. Testing, boreholes, geological and hydrogeological studies, etc., help to push the true cost of landfill operation up, but sometimes not to the levels which are actually charged.

## 4. SEA DISPOSAL

This embraces both the discharge to sea by way of direct pipelines or outfalls, and the use of ships to actually dump liquors, sludges (and even, in special cases, solids) into the sea. The former is the responsibility of the Water Authorities, the latter of MAFF (Ministry of Agriculture, Food and Fisheries). The former will be under some consent or agreement specifying conditions; the latter will be by specific licence issued only after serious consideration by MAFF under the Dumping at Sea Act, Britain's enabling legislation for the Oslo Convention.

Once again, sea disposal is not the route for the disposal of genuine solvent wastes, but is certainly a possible one for essentially aqueous wastes containing low levels of organic species such as solvents. I do not know, but would guess that hundreds of millions of gallons of effluent are discharged to the sea via pipeline each year, containing the equivalent of millions of gallons of solvent. The direct cost of this will be negligible if the cost of providing the pipeline or discharge outfall itself is disregarded.

Sea disposal from ships is a somewhat different proposition, requiring as it so obviously does, substantial hardware in the form of sea-going vessels and possibly on-shore bulk storage facilities. Nevertheless, this tends to be a high volume, fast turnaround business, with disposal charges themselves being well below the alternatives of incineration, biodegradation or whatever other physico-chemical techniques could theoretically be applied. The waste producer requires an annual licence at a nominal fee, and must supply certain analytical and other data as well as paying for any special tests which MAFF may want done.

The licensing procedure provides an effective screening system, and does result in some wastes being excluded from sea dumping. However, for the right sort of material, and given that a licence is issued, the sea dumping charge itself will be augmented by the cost of transporting the waste to the appropriate port. Overall, this quite often results in a reasonably economical option being available.

## 5. CATALYTIC OXIDATION

Recently, Re-Chem International examined the possibility of employing this technique for organic residue destruction. Our investigations very quickly convinced us that the technology had not yet been developed to the point where there was the slightest hope of it being used as a viable commercial approach for our type of application.

In fact, the technique is limited to emission control applications, i.e. gas streams emanating from some other process, and which contain very low levels of organic species. Even then, one must be extremely careful as to the nature of those organic species, as it is readily apparent that catalysts may easily be poisoned by organic species other than those for which they were designed.

Typical of the information which we received from the many catalyst manufacturers approached, was that around 1·5 cubic feet of catalyst would normally be required to handle a gas flow of up to 1000 standard cubic feet per minute, containing 1000–2000 ppm of suitable organic species. The gas stream requires heating to several hundred degrees centigrade, which, with the consequent risk of explosions involving the organic species present, imposes a further limitation on that concentration to one of not exceeding 10% of the lower explosive limit (LEL).

Such a system clearly has potential application for the purification of off-gas streams containing low levels of organic species. It will easily be seen that to contemplate such an approach for waste disposal, even assuming that suitable wastes and catalysts could be matched, would involve a catalyst bed size of enormous proportions. These catalysts are not cheap, and given the ease with which they can be poisoned, it is difficult to imagine a contractor considering this technique as being commercially viable.

The assumption that suitable wastes and catalysts could be matched is not, in itself, easily justified. Wastes are usually complex mixtures, given to considerable variation both in the ratio and in the nature of the materials contained. So many species are known to be problematical to catalyst life, that it will be difficult for the contract operator to find waste arisings which he is reasonably certain could be relied on not to vary beyond the acceptable limits.

## 6. CONCLUSIONS

Incineration is, and will probably remain, the single most effective means of destroying a wide variety of organic and organic-contaminated wastes. Sophisticated testing has demonstrated the high destruction efficiencies which can be obtained. However, incineration should never be thought of as a cheap disposal option because the cost of constructing suitable plant, and the subsequent operation of that plant and its peripheral business, will inevitably necessitate high disposal charges when compared with other disposal options, such as landfill.

## REFERENCES

1. R. B. Pojasek, *Toxic and Hazardous Waste Disposal*, Wiley, New York, 1978–80, 3 volumes.
2. R. A. Conway and R. D. Ross, *Handbook of Industrial Waste Disposal*, 2nd edn, Van Nostrand, New York, 1980.
3. International Technical Information Institute, *Toxic and Hazardous Industrial Chemicals: Safety Manual for Handling and Disposal with Toxicity and Hazard Data*, ITII, 1976.
4. DOE, *Waste Management Papers*, Nos 6, 9, 14 and 15, HMSO, London.

# PART IV

# HEALTH, LEGISLATION AND SAFETY

# 17

# LEGAL REQUIREMENTS CONCERNING THE USE OF SOLVENTS

M. R. WRIGHT

*Health and Safety Executive, Chelmsford, UK*

## ABSTRACT

*United Kingdom Health and Safety at Work Regulations endeavour to control the use, storage and transport of solvents in a working context in the light of two potential hazards: toxicity and flammability. Relevant Sections of, and Regulations made under the following Acts are discussed: Factories Act 1961, Offices Shops and Railway Premises Act 1963, The Petroleum Consolidation Act 1928, The Fire Precautions Act 1928, The Health and Safety at Work etc. Act 1974. Enforcing Authorities and the Health and Safety (Enforcing Authority) Regulations 1977 are reviewed.*

## 1. INTRODUCTION

At the 3rd Solvent Symposium for Industry in 1980, Mr J. A. H. Wood dealt with the comparisons between the traditional health and safety regulations of the United Kingdom which had developed from the first Factory Act of the early nineteenth century and the general approach which was adopted in the Health and Safety at Work etc. Act 1974.

In this paper it is my intention to summarise the present position of both types of legislation whilst inevitably touching on the differences, and because since 1980 there has been considerable legislative activity, to identify some of the areas where this activity is continuing and is likely to affect the use of solvents. It should however be made clear that much of what is to be written applies to many substances and not just solvents.

It should be noted that much of the 'traditional' legislation to which I have referred is still very much with us. A major provision of the Health and Safety at Work etc. Act 1974, and which is contained in Section 1, is the replacement of the 'traditional' legislation by new Regulations and Codes of Practice. This, however, must be seen as a very long-term objective and although the Act has now been in force for over eight years much work remains to be done.

Consequently, many legal peculiarities exist which present problems not only to the enforcing authorities but also to employers and employees. One example is a well-known Code of Regulations which will be referred to again—the Highly Flammable Liquids and Liquefied Petroleum Gases Regulations 1972. These are what I term to be 'traditional' legislation. They were made under the provisions of the Factories Act 1961 and Regulation 3 tells us quite specifically where those regulations apply. In short, they apply to factories as defined in the Factories Act and certain other places, such as warehouses and building sites, which are also subject to the Factories Act.

Consequently, many places of work which use highly flammable liquids, which include many solvents, are not, however, subject to the Factories Act. Hospital and University laboratories are but two examples. There the legal control for health and safety is exerted only by the general provisions of the Health and Safety at Work etc. Act 1974 and any Regulations and Codes of Practice which have been made under that Act.

## 2. POTENTIAL HAZARDS: TOXICITY AND FLAMMABILITY

As a general rule, solvents are not specifically referred to as such in either the 'traditional' or the 'new' legislation. They attract the provisions of such legislation simply by possessing certain characteristics which present certain hazards. In practice, Health and Safety legislation in this country identifies the potential hazards of solvents in two main areas—toxicity and flammability. The legislation therefore seeks to control these hazards in use, storage and transport.

In discussing the detailed requirements of various Acts and Regulations reference will be made to a number of Health and Safety Executive publications which are generally designed to give practical guidance and interpretation in respect of the legislation.

Four sections of the Factories Act 1961 and one of the Offices,

Shops and Railway Premises Act 1963 have or may have special relevance to the use or storage of solvents, either because of toxicity or flammability.

Section 4 of the Factories Act and Section 7 of the Offices, Shops and Railway Premises Act require effective and suitable provision to be made for the adequate ventilation of all workrooms and for rendering harmless fumes, dust and other impurities which may be injurious to health. Guidance Note EH22—'Ventilation of Buildings'— is available. Section 63 of the Factories Act 1961, on the other hand, requires more stringent precautions to be adopted to prevent, for example, toxic fumes entering the general atmosphere of a workroom in the first place. Such precautions include the enclosure of plant or the provision of effective local exhaust ventilation.

The publication by the Health and Safety Commission and Executive of information about toxic hazards is at present under review. For a number of years this has included the annual publication of a list of threshold limit values (TLVs) prepared by the American Conference of Governmental Hygienists. This was last carried out in 1981 with the issue of Guidance Note EH 15/80—Threshold Limit Values 1980.

The Health and Safety Commission which is guided by the Advisory Committee on Toxic Substances has now issued through the Executive Guidance Note EH 40 'Occupational Exposure Limits 1984'. The Note gives advice on limits to which exposure to airborne substances hazardous to health should be controlled in workplaces. The Note which lists occupational exposure limits and is divided into two parts, i.e. control limits and recommended limits, will be reprinted annually.

Control limits are contained in Regulations, approved Codes of Practice in European Community Directives or adopted by the Health and Safety Executive following detailed consideration of the available scientific and medical evidence. Such limits should not normally be exceeded. For example in 1983 a control limit in respect of trichloroethylene was adopted. The long-term exposure limit is set at 100 ppm (8 hour time weighted value) whilst the short-term exposure limit is 150 ppm (10 minute weighted value).

Recommended exposure limits are recommended to the Health and Safety Executive on advice from the Advisory Committee on Toxic Substances. These are considered to represent good practice and serve as realistic criteria for the control of exposure, plant design, engineering controls and, if necessary, the selection and use of personal protective equipment.

Whilst Section 63 is designed to protect employees from toxic hazards of solvents arising from a normal production process, Section 30 seeks to deal with the hazards which result from maintenance or cleaning operations. Where these require persons to enter a confined space such as a production vessel, where dangerous fumes are liable to be present, certain precautions must be taken. In general terms the space must be isolated, have been cleaned, ventilated and tested and found to be free from danger or alternatively the person entering the space must wear suitable breathing apparatus and use a lifeline. Guidance Note GS 5 gives practical guidance on these requirements.

Where solvents exhibit the hazard of flammability, or are potentially explosive, Section 31 of the Act may have relevance. This specifies precautions to be followed for the opening up of plant containing a flammable vapour under pressure and the measures to be followed where plant, tanks or vessels which may have contained flammable substances are to be subjected to the application of heat, such as welding. The Health and Safety at Work Booklet No. 32—'Repair of Drums and Small Tanks—Explosion and Fire Risks'—highlights some of the disastrous consequences of failure in this area and provides guidance on the necessary precautions.

## 3. CODES OF REGULATIONS

Turning to Codes of Regulations which have been made under the Factories Act 1961 and which relate to the use of solvents, both the Shipbuilding and Ship Repairing Regulations 1960 and the Construction (General Provisions) Regulations 1961 contain requirements which serve a similar purpose to that of Section 63 of the principal Act. The Shipbuilding and Ship Repairing Regulations do in addition contain in Regulation 49 a specific requirement about the use in confined spaces of flammable solvents in connection with the application or removal of paints. Published guidance about flammable hazards on building sites is available in Guidance Note EH 7—'Petroleum based Adhesives in Building Operations' and in Booklet HS(G)3—'Highly Flammable Materials on Construction Sites'.

Two further Codes of Regulations, this time dating from the 1920s—the India Rubber Regulations 1922 and the Chemical Works Regulations 1922—still remain in force and contain various requirements directed at very specific operations which include certain potential toxic hazards.

The Dry Cleaning Special Regulations 1949 are directed at a single operation which requires the use of solvents and is concerned with the hazard of flammability. The use of any liquid having a flash point below 90 °F (32 °C) is prohibited (except for spotting) for the cleaning of textile articles or wearing apparel.

A regulation which I think will have a much greater impact on industry in general than has been previously mentioned is Regulation 27 of the Electricity (Factories Act) Special Regulations 1908 and 1944. Although contained in the original 1908 Regulations it is in reality an example of how a generally worded requirement can remain legally relevant despite immense technical changes since drafting. The regulation 'simply' requires that all electrical conductors and apparatus exposed to explosive or flammable surroundings have to be 'so constructed or protected and such special precautions shall be taken as may be necessary adequately to prevent danger in view of such exposure or use'. I am sure that those who have to attend to such matters appreciate, if nothing else, that compliance with this Regulation can be extremely costly. Guidance from the Health and Safety Executive is contained in Form SH 928—'Memorandum on the Electricity Regulations'.

The last Code of the 'traditional' Regulations which will be referred to is the Code mentioned in Section 1 of this chapter—the Highly Flammable Liquids and Liquefied Petroleum Gases Regulations 1972. The use and storage of solvents are subject to these Regulations if the solvent concerned is a 'highly flammable liquid' as defined in Regulation 2. Basically these are liquids which give off a flammable vapour at a temperature of less than 32 °C and which support combustion. The precautions which the regulations require to be adopted include the provision of proper storage in safe positions, the taking of appropriate precautions against spills and leaks by conveying flammable liquids wherever possible within enclosed systems, the elimination of sources of ignition, the prevention of escape of vapours by enclosure and exhaust ventilation and the requirement that any enclosure within which flammable liquids are manipulated should be a fire-resisting structure.

## 4. FIRE PRECAUTIONS

I have already referred to publications relating to building sites. Other guidance information is contained in booklet HS(G) 4—'Highly

Flammable Liquids in the Paint Industry', and Guidance Note CS
1—'Industrial Use of Flammable Gas Detectors', CS 2—'The Storage
of Highly Flammable Liquids', PM 25—'Vehicle Finishing Units—Fire
and Explosions Hazards' and EH 9—'Spraying of Highly Flammable
Liquids'.

If highly flammable solvents are petroleum mixtures with a flash
point of below 23 °C storage arrangements are then subject to licensing
under the terms of the Petroleum (Consolidation) Act 1928. Administration of this Act is vested in County Councils.

Fire precautions in general, and by this I mean such matters as
means of escape from buildings, the provision of fire alarm systems and
the provision of fire fighting equipment, are certainly in the private
industrial sector normally the concern of the Fire Authority by virtue
of the Fire Precautions Act 1971. A guide to the application of this Act
in relation to factories is available, but in brief—a fire certificate is
required if more than certain numbers of persons are employed or if
explosive or highly flammable materials are stored or used.

The Fire Certificates (Special Premises) Regulations 1976, however,
identify some premises as having special risk, for example, the storage
of more than 4000 tonnes of highly flammable liquids, and places the
responsibility for the issue of the Fire Certificate on the Health and
Safety Executive. In the case of non-industrial premises the situation
may not be so straightforward and the only advice I can really give, if
doubt exists, is for an approach to be made either to the Fire Authority
or the Health and Safety Executive to ascertain whether a certificate is
required.

## 5. THE HEALTH AND SAFETY AT WORK etc. ACT 1974

The preceding sections of this chapter relate to the specific traditional
legislation which directly applies to industrial users of solvents. The
Health and Safety at Work etc. Act 1974, for which a guide is
available, applies almost universally to any work situation in this
country. Consequently, for the industrial organisation, both the
Factories Act 1961 together with its supporting Regulations and the
Health and Safety at Work etc. Act 1974 apply to its activities. For
establishments such as Universities and Colleges who may be large
users of flammable and toxic solvents only the provisions of the 1974
Act are legally relevant.

This has considerable legal implications. Returning to a part of the legislation already mentioned, Section 4 of the Factories Act 1961, we see examples of the more onerous duties imposed by the Factories Act than those in the Health and Safety at Work Act. Section 4 states: 'Effective and suitable provision *shall* be made for securing and maintaining adequate ventilation'. This is an absolute duty with which, once the area concerned is established as a workroom subject to the Factories Act, there is no valid legal excuse for not complying. The Section then continues to require the rendering harmless *so far as is practicable* of all such fumes, dust and other impurities which are generated. This is a less onerous duty than the absolute duty and if the matter to be remedied cannot be remedied in the light of current knowledge and invention, then that becomes valid legal defence.

Moving on to the general provisions of the Health and Safety at Work etc. Act 1974, and in particular to Section 2, we repeatedly see the words 'reasonably practicable'. For example, the provision concerning the 'arrangements for ensuring so far as is reasonably practicable safety and absence of risks to health in connection with the use, handling, storage and transport of articles and substances'. Thus not only do we have the defence regarding practicable we have in addition the word 'reasonably'. There have been many High Court cases which have determined the legal meaning of the combination of these words. They mean in essence that not only must the precaution be practicable, it has to be reasonable in light of the degree of hazard being considered and the costs involved. Section 40 of the Act places the burden of proving that it was not reasonably practicable to have done more than in fact was done on the accused. Having noted all this, as and when supporting regulations are made under the Health and Safety at Work Act, all the three levels of duty may be included.

The purpose of Section 2 of the Health and Safety at Work Act is, basically, to require all employers to examine their operations, identify potential hazards, devise appropriate precautions and fully advise and instruct their employees. Section 3 takes this duty a stage further and requires employers to consider from the Health and Safety viewpoint other persons such as contractors, visitors and persons living beyond the boundaries of premises who might be affected by their activities.

Section 6 of the Act placed duties on manufacturers and suppliers of, for example, solvents. In the case of machinery and plant the intention of that section is, of course, that such items are provided in safe conditions. Clearly, however, by their very nature solvents may be inherently dangerous because of either their flammability or toxicity or both. Here the aim of the Section is to ensure that proper guidance is given to users both about potential hazards and the precautions to be followed, that the solvents are in adequate containers and that testing and research work is carried out. Detailed guidance on these points is contained in Guidance Note GS 8—'Articles and Substances for Use at Work'.

Supporting the general provisions of Section 6 of the Health and Safety at Work Act there has emerged, and is still emerging a body of Regulations and Codes of Practice which both put flesh on these general provisions and which are putting into legal effect numerous Directives of the European Community.

## 6. REGULATIONS MADE UNDER THE HEALTH AND SAFETY AT WORK etc. ACT 1974

In 1978 The Packaging and Labelling of Dangerous Substances Regulations were made. It must be borne in mind that these regulations like most of those I have described are not specificially aimed at solvents. The Regulations proscribe a number of dangerous substances and impose requirements in respect of the containers in which they are supplied and about the marking and labelling. These Regulations have already been amended in 1981 by the addition of more substances. Further work is in progress with the intention of producing either comprehensive regulations or a set of regulations designed to cover the packaging and labelling of not only dangerous pure substances but also dangerous mixtures. Guidance on the existing Regulations is available in the form of Booklet HS(R) 1 and any proposed new regulations are or will be the subject of consultative documents.

Work is also in hand to have a comprehensive legal framework controlling the transportation of dangerous substances by road. Parts of the Dangerous Substances (Conveyance by Road in Road Tankers and Tank Containers) Regulations 1981 came into operation on 1 January 1982. Other parts came into operation on 1 January 1983 and the final part on 1 January 1984. These Regulations which revoked the Hazardous Substances (Labelling of Road Tankers) Regulations 1978 contain provisions relating to the construction and maintenance of vehicles and tank containers, the obtaining and giving of information and the marking of road tankers and tank containers. The Regulations are supported by an approved Code of Practice. Again guidance is available in the form of a Health and Safety Series Booklet HS(R) 13. Also in the same series is booklet HS(R) 10 (the approved list) containing approved substance identification numbers, emergency action codes and classifications for dangerous substances conveyed in road tankers and tank containers. Enforcement of thse Regulations depends upon the position at any moment in time of the vehicle concerned. On a road or in a public place the Police Authority has responsibility, whilst in a place of work the relevant enforcing authority exercises control.

A further Code of Regulations which came into operation in 1982 are the Petroleum Spirit (Plastic Containers) Regulations 1982, which permit the keeping of petroleum spirit in plastic containers for use as fuel for an internal combustion engine. An approved Code of Practice concerning the testing and marking of the containers has also been issued.

The Notification of New Substances Regulations 1982 also came into force in 1982. These again implement EEC Directives and are designed to ensure that the potential of new chemical substances to cause harm to man or to the natural environment is assessed before they are placed on the European Community Market. Details about the substances within the scope of the Regulations and the notification procedures are given in booklet HS(R) 14.

It has already been mentioned that work is in hand regarding regulations for packaging and labelling. Similar action is being pursued for road transportation of dangerous substances other than by tankers. Also the whole field of the use of highly flammable liquids is under review so that perhaps by the time of the next Symposium we may have new regulations in this matter or at least consultative publications. Work also continues on regulations dealing with pressurised systems and on a replacement of the existing Electricity Regulations.

## 7. ENFORCING AUTHORITIES

Enforcement of the Factories Act, the Offices, Shops and Railway Premises Act and the Health and Safety at Work etc. Act is allocated between the Health and Safety Executive and Local Authorities generally in accordance with the Health and Safety (Enforcing Authority) Regulations 1977. Broadly, local authorities—that is District Councils as opposed to County Councils—enforce in such places as commercial offices, banks, shops and wholesale warehouses. The remainder which includes factories, universities, schools and hospitals are dealt with by the Health and Safety Executive. The powers of Inspectors are specified in Sections 20–23 of the Health and Safety at Work Act, but I would just like to highlight two powers—that to take samples and that to issue Notices.

In the case of the former, where for example solvents are used or stored, an Inspector may take samples with the general intention of having the sample analysed—probably to determine the flashpoint or chemical composition—so as to be able to come to a conclusion about toxicity. The results of the sampling may be used in any subsequent proceedings.

Secondly, the power to issue Notices. These may take the form of Improvement or Prohibition Notices. Improvement Notices require certain action to be taken within a specified period, for example, improvements to a highly flammable liquids store. A Prohibition Notice which may either take immediate or deferred effect is designed to stop an unsafe practice and prohibit its re-starting until certain remedial measures are taken. For example, the operation of an evaporating oven without explosion relief. Guidance on such ovens incidentally is given in booklet HS(G) 16.

Both types of Notices are subject to the appeals procedure set out in the Act. Such appeals are determined at an Industrial Tribunal which has powers to set aside, amend or confirm the Notice.

Finally, of course, the Act contains provisions about contraventions of the legislation, 'new' or otherwise. Section 33 which has been amended by the Criminal Law Act of 1977 provides penalties of up to £1000 on summary conviction whilst on indictment the financial penalty is unlimited and there is also the possibility of imprisonment for up to 2 years.

# 18

## HEALTH ASPECTS OF TOXIC SUBSTANCES: A TRADE UNION VIEW

D. Gee

*General and Municipal Workers' Union, Esher, Surrey, UK*

## ABSTRACT

*It is essential to assume that toxic substances are more likely to be harmful than harmless and to consider all available types of evidence. The latter should be weighed against the 'balance of probabilities' level of proof. It is recommended that risks and benefits should be assessed by an administrative procedure that includes all interested parties and that control limits should be based on the exposure levels that a majority of users can achieve. Substances should be banned, authorised or controlled depending on the risks and costs in each case. Broadly similar control limits should apply to industries and substances in all countries.*

## 1. INTRODUCTION

The trade union movement has recently campaigned for a ban on the use of the herbicide 2,4,5-T and benzidine-based dyes and for strict controls on the use of glass fibre and vinyl chloride. We are also demanding much lower control limits than the Health and Safety Executive (HSE) currently recommends on scores of other materials. What is the basis for these policies? Is it an hysterical response to horror stories from the media, or a serious attempt to avoid history repeating itself? Let us look at how toxic substances were dealt with in the past to see if it provides any explanation for today's attitudes.

The old approach was to assume substances were safe until there was overwhelming human evidence that could not be explained away, then to introduce ineffective legal controls some years later, and finally to tighten up the sanctions when it was economically easy to do so. The result was what we call 'The Generation Game' which has been played with the lives of those workers forced to gamble with someone else's materials. Unfortunately for them, the people who made decisions about the safety of materials did not run the risks themselves. Had they done so, the game might have been shorter. This is how it is played. One generation of workers is exposed to a new material, which is assumed to be safe. (This happens even when previous generations in nearby countries have already had their generation game, as was the case with $\beta$-naphthylamine in dyestuffs' manufacture before and after the First World War.) At the end of this exposure period some workers are beginning to identify a hazard, like the asbestos textile workers who complained in 1900, or the Sheffield grinders who sang about their dusty plight in the 1860s. By the end of the second generation, a radical doctor or scientist is pointing the finger of suspicion at the culprit material, but the medical/scientific establishment of the day shout him down. At the end of the third generation the evidence is becoming overwhelming and the speed with which action is taken is then determined solely by economic considerations. If controlling or eliminating the substance is going to hurt business then 'defensive science' and political opposition can be invoked for perhaps another generation or two. Controls are finally accepted when the alternative materials or processes are fairly easy to adopt. By that time several generations of victims can testify that the 'innocent until proved guilty' approach to workplace substances is expensive for them. All of the major toxic substances have been through variations of this 'Generation Game', from silica and mule spinning oil, to $\beta$-naphthylamine and asbestos. So what is our alternative approach?

## 2. TOXIC SUBSTANCES ARE LIKELY TO BE MORE HARMFUL THAN HARMLESS

Beginning with the assumption that substances are more likely to be harmful than harmless, it follows that they should be handled with care and caution. (It can be argued that the Health and Safety Executive (HSE) have adopted this approach in their Guidance Note EH. 18 on 'Toxic Substances: A Precautionary Policy'.) Several important implications flow from this initial assumption.

First, unnecessary exposures to substances are avoided. Getting covered in dust as part of the general attempt to prove or assume safety (as was done with asbestos, silica and the other known killers) will not be a necessary part of the job, giving any future killer materials less of a heyday. Secondly, control technology (ventilation systems, process controls, respirators, etc.) will be more effective and cheaper as demand for it increases. Instead of one small part of industry at a time having to install and often invent appropriate equipment when one substance is indicted, all industries would be obliged to use at least some control technology. Its price would fall, and savings would be made in raw materials and in throughput efficiency. The relative cost of getting a future killer substance to zero exposure would also fall. Thirdly, pre-market testing would be mandatory for all new substances used at work. Worker guinea pigs would no longer be acceptable as post-market test material. And finally, one or two negative findings in experimental tests would not invalidate the initial assumption of harm. This would reverse the current position where industry does not accept one or two positive findings (despite the statistical difference in significance) as invalidating their initial assumption that materials are safe.

## 3. CONSIDERING ALL AVAILABLE EVIDENCE OF HARM

Our new approach to substances then goes on to consider all classes of evidence of harm, giving weight to positive examples from all of them, and discounting many of the negative findings on methodological grounds. Thus, good experimental evidence from (i) bacteriological systems (the Ames test, etc.), (ii) animal experiments, (iii) similarity of chemical or physical structure, and (iv) workers' experiences, are given as much weight as rigorous epidemiology. Waiting for one or two generations to identify cancers is not now acceptable.

In the past negative epidemiology has been the main barrier to earlier control. This took the form of either specific studies that purported to show that a substance was in fact safe despite circumstantial or animal data to the contrary, or it was merely the common-sense 'look-at-my-workforce-they're-all-there' approach which was even less valid, if superficially convincing. The combination of the 'healthy worker' and 'survivor population' effects was difficult to overcome. Now we are saying that negative epidemiology is unconvincing, particularly in the face of positive *in vitro* and animal data. It is reassuring to see several eminent 'experts' including ones from ICI, supporting OSHA's similar stand.

The available evidence is then weighed against a 'balance of probabilities' level of proof, rather than 'beyond all reasonable doubt' of scientific method. Substances should not be given the protection of the criminal law defence but at most should be judged by civil law criteria, as is already the case with radiation. This would eliminate the usual definition of scientific proof from occupational health. Waiting for it in the past has been a costly affair, at least in human terms.

## 4. RISKS AND BENEFITS

The next stage in the approach is to involve all interested parties in the essentially political process of considering the risks and benefits of using and controlling substances at work. In the past, 'science' has enveloped and mystified the 'politics' within decision making, and human guinea pigs have been left out altogether. Workers and their representatives need to be involved in the process of hazard and risk assessment because: (i) it is morally indefensible for those who have to run the risks of, for example, occupational cancer to be excluded from the process of deciding the size of those risks; and (ii) the process of hazard and risk assessment involves value judgements, opinions, economic and political choices, as well as scientific fact and judgement. For example, there is a choice to be made at the outset of hazard assessment—Is an unknown substance to be judged innocent of toxic potential until proved guilty, or guilty until proved innocent by years of well-researched use? Most employers assume the former, while unions assume the latter. Another choice is between negative epidemiological evidence and good positive experimental evidence—many company doctors prefer the former, whilst unions prefer the latter.

Value judgements are also involved in decisions about strength of evidence for a particular association and more importantly, in deciding what level of proof is sufficient to classify, for example, a carcinogen. The usual scientific level of proof, which is similar to that used in the criminal courts, i.e. that the evidence should prove guilt 'beyond all reasonable doubt' is inappropriate for assessing evidence of toxic effect, particularly where the effect is carcinogenic. The 'cost' of guilty substances going free from rigorous regulation is suffering and death amongst workers. Many authorities now agree with the unions that a lower level of proof, i.e. the 'balance of probabilities' used in the civil courts, is more appropriate for assessing evidence of toxic effect. The 'cost' of condemning an innocent substance is a marginal misallocation of some resources, where perhaps less stringent controls would have been acceptable. (This potential cost becomes less and less as employers begin to adopt the HSE's 'precautionary policy' of treating all substances as though they were potentially toxic. See HSE Guidance Note EH. 18.)

Hazard assessment also involves a *knowledge of actual user conditions.* Workers have a long and painful experience of feeling the difference between how production and work systems should have been organised and what actually happens. Any realistic hazard assessment needs to incorporate views about *the practice* of controlling substances at work, as opposed to *the theory,* and workers can provide unique experience of this. Although it is true in theory that effective control measures can contain the most toxic material, this does not happen in practice. Workers' experience is needed to counter the technological arrogance of some employers and scientists in determining the ease with which substances can be controlled.

Workers must also be given the chance to determine the risks that they will be asked to run. Up until recently, workers had no say in determining 'acceptable risk'—it was decided for them by scientists and policy-makers. The tripartite structures of HSE now enable some representation of workers' views, but the unions need to have more access to sympathetic expertise, the proceedings should involve statements from interested parties and their expert witnesses, and the evidence and arguments used should be freely available for public and peer group scrutiny.

Pesticide safety remains anachronistic and undemocratic. The pesticide 2,4,5-T, for example, was cleared as 'safe' by a committee of scientists and administrators who will never use a back spray in difficult winds in their lives. Trade unionists are now demanding a say in the risks they are asked to run, and are putting a different kind of price on the risks they will accept. For example, the benefits to society of a new substance should be clearly demonstrated before it is considered for marketing, and at least part of the costs of health risks should be borne by the risk-making industry, and not left to individual workers and their families, or to the NHS and the general community.

Long-term health hazards can cost the offending industry very little in the present system, as the cost of crippled workers in the future is usually externalised. Compensation for harmed individuals is still available to only a minority of the victims of known killers. Automatic, no-fault insurance schemes should cover all present and future victims of harmful substances, so that any gain to 'society' is not balanced against a risk that only a non-random few are asked to bear alone. Money and economic security cannot replace health, but it helps. The industry that pollutes or cripples should bear the full economic costs of any future killer so that individuals do not suffer, and industries and substances are not subsidised. At present many asbestos victims get no compensation for pain, death and economic disruption, and the community cost of the 'magic mineral' is enormous. The overall cost of ill-health and of dealing with old lagging probably makes asbestos the most expensive building material ever used. The 2000 mesothelioma cases since 1966 represent over £70 million on one calculation; and if stripping one small power station several years ago cost over £1 million, how much is the clean up of housing estates, schools, hospitals and factories now going to cost? The risk/benefit assessment for present materials must take account of likely life-cycle costs if future generations are not to be burdened with our expensive mistakes.

## 5. SETTING CONTROL LIMITS

The trade union approach to the setting of *control limits* is also different from that of the employers and regulatory authorities. *First*, for carcinogenic substances, we do not accept that there is a known safe level at all, so that any exposure involves an increased risk of cancer, albeit small in some cases. *Secondly,* we do not accept the use of risk comparisons in other industries to set risk levels in one particular industry, because that would give undue weight to the most dirty and dangerous industries. They are bad enough on their own without dictating the pace of improvements in other industries. (This comparative 'index of harm' approach has been adopted by the ICRP (International Commission on Radiological Protection) for setting radiation control limits, but rejected by the Simpson Report on Asbestos in the UK—see para. 162 of Vol. 1.) *Thirdly*, we want to see control limits based on what the majority of a user group (industry, or parts of many industries, etc.) can achieve in the circumstances, with the 'dirty end' of the user group being given temporary exemptions to work to higher control limits. The opposite usually happens at present, where the 'dirty end' of a user group dictates the control limit, with the majority of the user group being expected to get 'as low as reasonably practicable' ('ALARP'). This does not happen in practice, because everyone works to figures, not moral dictums, and ALARP is not enforceable by either government inspectors or safety reps. The present practice condemns the majority of workers in a user group (possibly up to 95% of them) to higher risks than necessary, and allows a minority of less responsible employers to dictate risks in the rest of their industry. The only serious defence for this procedure is that it imposes the same risks on all workers, whereas our approach would lead to differential risks. The only authoritative source we know of that has considered this question, the Simpson Report on Asbestos, ruled out our approach on the grounds that it would mean that some risk levels in some processes are more acceptable than others, 'an impression which we would not wish to create' (para. 164 of Vol. 1).

Differential risks are a fact of life. The risks and benefits available to workers at the dirty end of a user group in an area of high unemployment are very different from those facing workers at the clean end of the same user group in an economically prosperous region. For example, the position of asbestos textile workers in the North of England is very different from asbestos cement workers in the South of England. On current evidence a 1 fibre/cm$^3$ asbestos control limit, which is likely to be associated with a 1 or 2% excess death rate from asbestos diseases, is achievable in the textile factory, but 0·2 fibre/cm$^3$ with a correspondingly lower excess death rate is achievable in the cement factory.

There is no good reason why the asbestos cement worker should run unnecessarily high risks because of the problems of the asbestos textile factory. All of us accept differential risks as part of life depending on where we live, how we travel and what we eat, so occupational life need be no different, so long as the different risks are clearly explained to those expected to run them. A final bonus to our approach would be that the publication of different control limits for the same substance would increase the pressure on the 'dirty end' of a user group to clean up because workers would want to achieve the lower risks of the clean end as soon as possible. In the present situation there is very little publicity about the exposure levels that the clean end of an industry is achieving, and therefore less pressure for improvement.

There is not much disagreement with employers and enforcing authorities on *the steps involved* in setting control limits, but it is worthwhile summarising them:

(i) Assess the medical and epidemiological evidence and attempt to derive dose limits that would generate either no effect, or given risks of disease, depending on the substance, e.g. the dose limits associated with 0·01%, 0·1%, 1% and 10% excess disease rates amongst exposed workers.

(ii) In the absence of human dose data that could predict these dose/effect relationships, which is the usual case, use animal data with sufficiently large safety margins, e.g. 1000 times lower than the 'no effect' limit, to estimate the 'safe' limit for non-carcinogenic but toxic substances. This limit is to be preferred to all but the 0·01% excess disease limit from human data if it is lower.

(iii) Assess the exposure levels currently achieved in the user groups by appropriate monitoring surveys.

(iv) Assess the costs of achieving what are considered to be 'acceptable' levels of risk for the exposed workers in the user group. (In practice this 'acceptable' risk level varies from the 0·01% excess deaths claimed by ICRP for radiation control limits to over 10% excess byssinosis cases at the UK cotton dust standard.)

(v) Agree control limits on the basis of a compromise between these health, exposure and economic data, and publish the evidence and arguments used in the process of determining the limit.

## 6. INTERNATIONAL ENFORCEMENT

The next stage in the new approach is to demand maximum publicity for the health risks and broadly similar control limits, and their international enforcement, for substances in all countries so that we avoid the present export of hazardous materials and industries to developing countries. Reasonable US control of asbestos led to its flight to Mexico, and subsequent import back into the USA. Dyestuffs had a similar fate, and the way is open for many other industries to exploit the weaker position of developing countries. This is perhaps the most difficult part of the new approach because it involves international action and solidarity, the possibility of import controls in the countries with the best standards, and an underlying assumption that our risk/benefit choices should be broadly similar for everyone. This clearly is not so, as our argument for differential control limits in one country has shown. The differences between countries will often be greater than within countries. If life expectancy is 35, then cancer probabilities assume less importance than when life expectancy is 70. Without this kind of international co-operation, however, the most desperate and exploited countries will not only remain so, but will dictate the pace of improvement in working conditions elsewhere.

The final stage in the new approach is enforcement of the improved health standards by Workers' Inspectors appointed by the unions from amongst those who run the risks. It is only through trained and active safety representatives, backed up by HSE inspectors and specialists, that workplace improvements can be achieved and maintained. There must be no repetition of the asbestos story, where information about its danger to health was somehow forgotten or ignored for over 50 years.

## 7. NEW APPROACH

What happens when we apply this new approach to a number of substances currently in the news? The short answer with 'man-made mineral fibre' (MMMF) is that you get the first minority report in an HSE health hazard committee since the tripartite committees began in 1975. The majority report in the recent HSE Discussion Document on 'Man-Made Mineral Fibre' said that the substance should be treated with 'suspicion' and recommended two new control limits: 5 fibres per cm$^3$ and 5 mg per m$^3$. The TUC minority report to which we were party said that 1 fibre per cm$^3$ was needed, but we would compromise at 3 fibres per cm$^3$ if the employers would accept it. They did not, so our own position has reverted to 0·5 fibres per cm$^3$, the same as our initial target for asbestos. How do we arrive at that conclusion? If we apply our new approach we find that:

    (i) MMMF is assumed to be harmful at the outset.
   (ii) There have been few good studies of workers exposed to MMMF and none at all of workers who use, as opposed to manufacture, MMMF.
  (iii) Only a couple of epidemiological studies had a sufficiently long follow up to merit review and both these showed some excess of respiratory disease, with one sub-group thought to have

been exposed to respirable fibres showing an excess of malignant and non-malignant respiratory disease.

(iv) It is the size and shape of asbestos fibres which determine its carcinogenicity, and MMMF is now produced in vast quantities containing a substantial proportion of fibres within the critical size range of 3 μm and less in width and 5 μm or more in length.

(v) This respirable fibre range has only been in *mass* production since the 1960s, so we do not expect any clear human evidence of cancer from it until the 1990s.

(vi) MMMF does not seem to split longitudinally into fibrils, as asbestos does, so it may not be as potent as asbestos.

(vii) The cancers in a Turkish village (called the equivalent of 'pain') which seem to be caused by natural fibres other than asbestos are more evidence that fibres of the critical size and shape are sufficient to cause cancer.

(viii) Animal experiments with implanted MMMF cause tumours in the same way as asbestos; inhalation experiments are negative so far but that is likely to be a question of technique.

(ix) Present exposures in the MMMF manufacturing plants are generally below 1 fibre per cm$^3$, so a 0·5 fibre per cm$^3$ standard can be achieved without much economic cost. The 'dirty' exception to a low fibre count standard is the ceramic fibres plant, and its equivalent on the gravimetric standard is the construction site. Neither should be allowed to dismiss the evidence that supports the control limits we demand.

Applying our approach to vinyl chloride leads us to demand a 1 ppm standard for vinyl chloride and that employers should recognise that it causes cancer of the brain and lung, and probably the blood-forming system, as well as liver cancer, AOL (acro-osteolysis) and other non-specific diseases like those for which our Vinatex members recently got £¾ million in compensation. If the past approach to toxic substances is applied to the question of whether VCM has reproductive effects the answer is no, as was recently declared by a company doctor with interests at stake. If we apply our approach the answer is a convincing yes, based on positive *in vitro* and animal data, chromosome damage in vinyl chloride workers, positive epidemiology from the USA and very unconvincing negative epidemiology.

What about the herbicide 2,4,5-T? The union approach to toxic substances has clearly established the case for a ban. The *in vitro* and animal data, taken in conjunction with the circumstantial and scientific human evidence, is sufficient to establish the probability of harm. The TUC decision to ban, as opposed to more rigorous control of use, was taken after assessing the numbers at risk, the scant benefits of continued use, and the availability of substitutes.

Finally, the TUC is calling for a ban on benzidine-based dyes because of the probability that they are carcinogenic and substitutes are available. The American factory inspectorate, OSHA, has an approach to toxic substances which is much closer to ours than that of the HSE, and their respective conclusions about benzidine dyes illustrate the difference. In January 1980 the HSE's Advisory Committee on Toxic Substances concluded that a carcinogenic risk 'could not be excluded', whilst OSHA, on the basis of the same scientific evidence, concluded that the dyes should be regarded and handled as carcinogens and replaced wherever possible. OSHA also took the wise step of extending their concern to the related *o*-toluidine and *o*-dianisidine dyes which are also likely to be carcinogenic.

The four examples above illustrate what our approach to substances means in practice. There will be an increasing necessity to apply the approach to many more substances as occupational health becomes more important than safety. General regulations on toxic substances, similar to the USA legislation, are clearly necessary if thousands of substances are to be dealt with effectively. Separate regulations on carcinogens are also essential. Health hazards are already more important in terms of the numbers of occupational deaths, disease and sickness compared to fatal accidents and injuries. Unless employers and health authorities in this country come much nearer to our approach to toxic substances, then there will be conflict over when and how they should be controlled. The 'generation game', we hope, is over.

# 19

# EEC DIRECTIVES FOR LABELLING OF SOLVENTS

S. Fumero* and G. Mosselmans

*Directorate for Internal Market and Industrial Affairs, Commission of the European Communities, Brussels, Belgium*

and

A. Berlin

*Health and Safety Directorate, Commission of the European Communities, Luxembourg*

## ABSTRACT

*This chapter reviews the approach made by the European Community from 1960 up to the present relating to the legislation concerning dangerous substances and preparations, and especially solvents. An analysis is made of the different aspects of the Community legislation covered by the following directives:*

— *Directive 67/548/EEC relating to the classification, labelling and packaging of dangerous substances.*
— *Directive 73/173/EEC relating to the classification, labelling and packaging of dangerous preparations—solvents.*
— *Directive 80/1107/EEC on the protection of workers from the risks related to exposure to chemical, physical and biological agents at work.*

## 1. INTRODUCTION

The control of chemicals, not specifically solvents, has been approached through two different and almost independent routes at Community level.

* Present address: Istituto de Ricerche Biomediche 'A. Marxer' RBM, Ivrea, Italy.

The first route had as its goal the elimination of technical barriers to the intra-community trade: this goal could only be reached by a unified approach towards the classification, packaging and labelling of the chemical substances.

Two principal directives were then elaborated within the framework of the Community:

— the 1967 Directive on the classification, packaging and labelling of dangerous substances and especially its Sixth Amendment (1979);[1,2]

— the 1973 Directive on the classification, packaging and labelling of dangerous preparations (solvents) and its 1980 Amendment.[3,4]

Although the programme for the elimination of technical barriers excludes any guidance about the use of the substances in their relevant fields of application, nevertheless their classification with respect to danger for man, the specifications of the hazards and the provision of safe-handling advice, furnishes indirect but important elements for the protection of the workers and the public at large.

The second route has as its goal the improvement of the safety and the health of the workers: for this goal an action programme on Safety and Health at Work, places special emphasis on the danger of exposure to chemicals: six of the fourteen actions of the programme call for harmonisation of limit values, setting of hygiene measures, developing criteria and biological indicators.[5]

The Council Directive of 27 November 1980 on the protection of workers against the risks connected with an exposure to chemical, physical and biological agents at work, provides the first direct element for active measures to be taken at Community level for the protection of workers against the harmful effects of certain solvents.[6]

## 2. SIXTH AMENDMENT

The directive 79/831/EEC, better known as the Sixth Amendment, introduces some important concepts and procedures which will now be considered.

### 2.1. Premarketing Notification

Before putting a *new* chemical on the market (*new* means not included in the European inventory) the chemical industry must send a notification dossier to the competent authority of the Member State where the chemical is to be commercialised. This notification dossier must include:

(a) A technical dossier supplying the information necessary to evaluate the risks which the new substance may entail for man and the environment. It should contain at least the information and results of the experimental studies referred to in the so-called base set (Annex VII) which concerns physicochemical, toxicological and ecotoxicological tests; the following list shows the toxicity tests as foreseen in the base set:

1. $LD_{50}$ oral, inhalation, cutaneous (usually two routes of administration);
2. skin irritation;
3. eye irritation;
4. skin sensitisation;
5. subacute toxicity;
6. mutagenicity (one bacterial and one non-bacterial test);

Further toxicological tests may be required in the case of chemicals produced in quantities of more than 10 tons per year.

(b) A declaration concerning the unfavourable effects of the substance in terms of the various uses envisaged.

(c) The proposed classification and labelling.

(d) The proposals for any recommended precautions relating to the use of the substance.

## 2.2. Concept of Dangerous Substance

Fourteen definitions of dangers (health and physical dangers) are given in the directive. Of these 14 types of dangers, nine are covered by symbols. The following list reports definitions of dangers:

1. explosive
2. oxidising
3. extremely flammable
4. highly flammable
5. flammable

6. very toxic
7. toxic
8. harmful
9. corrosive
10. irritant
11. dangerous for the environment
12. carcinogenic
13. teratogenic
14. mutagenic

## 2.3. Labelling Requirements

The EEC label is intended to give information on two types of dangers: health and physical dangers. The EEC labelling requirements are intended to provide a clear primary means by which all persons (workers as well as public at large) handling or using substances are given essential information about the inherent dangers of certain such materials. The means used are a combination of symbols, standard risk phrases (R-phrases) and standard safety advices (S-phrases). The symbols highlight the most severe hazards presented by the substance; the R-phrases try to give a more specific picture of those hazards, and the S-phrases give safety advice on necessary precautions and/or of mishandling to be avoided relating to the use of the substance.

In particular, the substances shall be classified as very toxic, toxic or harmful according to the following criteria:

(a) acute toxicity tests as shown in Table 1.

TABLE 1

| Classification | $LD_{50}$ oral (mg/kg) | $LD_{50}$ cut. (mg/kg) | $LD_{50}$ inhal. (mg/litre/h) |
|---|---|---|---|
| Very toxic | $\leqslant 25$ | $\leqslant 50$ | $\leqslant 0 \cdot 5$ |
| Toxic | 25–200 | 50–400 | $0 \cdot 5$–2 |
| Harmful | 200–2000 | 400–2000 | 2–20 |

(b) if data show: that for the purpose of classification it is inadvisable to use the $LD_{50}$ or $LC_{50}$ values as the principal basis because the substances produce other effects;
    or because of the existence of effects other than the acute effects indicated by experiments with animals, e.g. carcinogenic,

mutagenic, allergic, subacute or chronic effects, substances shall be classified according to the magnitude of their effects.

The basic concept expressed in the directive for the classification of these substances is that the chemical is classified on the basis of the highest degree of danger, not on the additivity of the different effects induced by the substance.

Naturally, for existing chemicals, classification and labelling must take place in so far as the manufacturer may reasonably be expected to be aware of their dangerous properties. For new substances, classification and labelling is mandatory and will be based on the data submitted to the proper authority in the notification dossier.

## 3. SOLVENTS DIRECTIVE

Up to now the toxicity and classification of substances was only considered when they were present in the pure form. Solvents are mostly used as mixtures of variable composition. To cover this aspect of classification and labelling a Council Directive on preparations was first adopted in 1973 and subsequently modified on several occasions.

For example, to determine if a preparation is to be classified and labelled as toxic or harmful, an empirical computation system has been set up based on a classification index for toxic and harmful substances which may grade this preparation. The indexes to be used in this classification are summarised in Table 2.

TABLE 2

Classification Indexes for Toxic and Harmful Substances for the Calculation of the Toxicity of Preparations

| Class of substance | Classification index $I_1$ | Exemption index $I_2$ |
|---|---|---|
| Very toxic and toxic I/A | 500 | 500 |
| I/B | 100 | 100 |
| I/C | 25 | 25 |
| Harmful II/A | 5 | 20 |
| II/B | 2 | 8 |
| II/C | 1 | 4 |
| II/D | 0·5 | 2 |

The equations to be used for the computations are given in Table 3.

TABLE 3
Equations for the Calculation of Toxicity of Preparations

| (1) | $\Sigma [P \times I_1] > 500$ | Toxic |
|-----|-------------------------------|-------|
| (2) | and $\left.\begin{array}{l} \Sigma [P \times I_1] \leqq 500 \\ \Sigma [P \times I_2] > 100 \end{array}\right\}$ | Harmful |
| (3) | $\Sigma [P \times I_2] < 100$ | No classification |

$P$ = Per cent by weight of each substance.
$P_{min}$: class I < 0·2%; class II < 1%.

The solvents have been graded in seven categories. To each category corresponds an index of classification and one index of exemption. The solvent preparation is classified on the basis of the result obtained by applying a simple equation whose terms are the percentages of the solvents and the corresponding indexes.

All substances classified or present in Category I are listed in Table 4.

TABLE 4
Substances Classified in Category I (1982) for the Calculation of Toxicity of Preparations

| I/A | I/B | I/C |
|-----|-----|-----|
| Carbon disulphide | Bis(2-chloro- | 1-Bromopropane |
| *Benzene | ethyl) ether | Methanol |
| Carbon tetrachloride | Phenol | Acetonitrile |
| 1,1,2,2-Tetrachloroethane | Cresol | 2-Hexanone |
| Pentachloroethane | Furfural | |
| Nitrobenzene | *Piperidine | |
| Aniline | | |
| *1,1,2,2-Tetrabromoethane | | |
| *Allyl alcohol | | |
| *1,2-Dibromoethane | | |
| *2-Chloroethanol | | |

Examples of the use of these indexes for the classification of preparations are given in Table 5. In addition, of course, the manufacturer has to indicate if the preparation presents other dangers, such as flammability or corrosivity, and give safe-handling advice.

It must however be emphasised that these directives *per se* cannot

TABLE 5
Examples of Classification of Solvents

| | 1,1,2,2-Tetrachloroethane | 1,1,2 Tri-chloroethane | 1,2-Dichloro-ethane | Cyclo-hexanol | $\Sigma (PI)$ | Classification |
|---|---|---|---|---|---|---|
| $I_1$ | 500 | 5 | 2 | 0·5 | | |
| $I_2$ | 500 | 20 | 8 | 2 | | |
| % Weight composition (P) (1) | 1% | 9% | 10% | 80% | $\Sigma PI_1 = 605$ | Toxic |
| (2) | 0·5% | 0·5% | 1·0% | 98% | $\Sigma PI_1 = 300$ $\Sigma PI_2 = 464$ | Harmful |
| (3) | — | — | 2% | 98% | $\Sigma PI_1 = 51$ $\Sigma PI_2 = 212$ | Harmful |
| (4) | (0·1%) | (0·5%) | 9% | 10% | $\Sigma PI_1 = 23$ $\Sigma PI_2 = 92$ | No classification |
| (5) | 0·3% | (0·5%) | 1·0% | 10% | $\Sigma PI_1 = 157$ $\Sigma PI_2 = 178$ | Harmful |

formally contribute to the worker protection since they do not and cannot lead to any prohibition or limitation of use at work.

## 4. PROTECTION OF WORKERS FROM CHEMICAL DANGERS AT WORK

The 1980 Council Directive[6] is a broad Framework Directive for the protection of workers from chemical dangers. This should result in all Member States following a similar path in the future. This Directive sets out two objectives:

- eliminate or limit exposure to chemical, physical and biological agents and prevent risks to workers' health and safety;
- protect workers who are likely to be exposed to these agents.

This Directive which will affect the majority of workers in the Community requires the Member States to take short-term and longer-term measures. The short-term measures, not applicable to solvents, require that, within three years, workers and/or their representatives shall have access to appropriate information concerning asbestos, arsenic, cadmium, lead and mercury, and within four years appropriate surveillance of the health of workers exposed to asbestos and lead shall take place.

The longer-term measures apply when a Member State adopts provisions concerning an agent. In order that the exposure of workers to agents is avoided, or kept at as low level as is reasonably practicable, Member States must comply with a set of requirements and inform the Commission, but in doing so they have to determine whether and to what extent each of these requirements is applicable to the agent concerned. These requirements are the following:

- limitation of the use of the agents at the place of work;
- limitation of the number of workers exposed or likely to be exposed;
- prevention by engineering control;
- establishment of limit values and of sampling procedures, measuring procedures and procedures for evaluating results;
- protection measures involving the application of suitable working procedures and methods;
- collective protection measures;

— individual protection measures, where exposure cannot be avoided by some other means;
— hygiene measures;
— information for workers on the potential risks to which they are exposed, on the technical preventive measures to be observed by workers and the precautions taken by the employer and to be taken by workers;
— use of warning and safety signs;
— surveillance of the workers' health;
— keeping updated records of exposure levels, lists of workers exposed and medical records;
— emergency measures for abnormal exposures;
— if necessary, prohibition of part or all of the agent, or agents, involved, in cases where use of the other means available does not make it possible to ensure adequate protection.

In addition, an initial list of eleven agents, including four solvents (Table 6), relates to the implementation of further, more specific requirements, which are as follows:

— providing medical surveillance of workers by a doctor prior to exposure and thereafter at regular intervals. In special cases, it shall be ensured that a suitable form of medical surveillance is available to workers who have been exposed to the agent after exposure has ceased;
— access by workers and/or their representatives at the workplace to the results of exposure measurements and, in the case of an

TABLE 6

Solvents Requiring Individual Directives in Accordance with the 1980 Council Directive[6] on the Protection of Workers

| Solvent | 1973 Solvents Directive (and Amendment) Classification | IARC (1982) Classification |
|---|---|---|
| Benzene | I/A | 1 |
| Carbon tetrachloride | I/A | 2B |
| Chloroform | II/A | 2B |
| p-Dichlorobenzene | — | Insufficient data for evaluation |

agent for which such tests are laid down, to the anonymous collective results of the biological tests indicating exposures;
— access by each worker concerned to the results of his own biological tests indicating exposure;
— informing workers and/or their representatives at the workplace where the limit values are exceeded, of the causes thereof and of the measures taken, or to be taken, in order to rectify the situation;
— access by workers and/or their representatives at the workplace to appropriate information to improve their knowledge of the danger to which they are exposed.

This Directive also requires the Member States to consult the social partners when the above requirements are being established.

With regard to the eleven agents mentioned above, the Council on proposal of the Commission will fix in individual Directives the limit value(s), as well as other rules, like those on lead and asbestos, now being discussed by the Council. Special emphasis will be placed, where possible, on the basis of scientific information and analytical practicability on the biological monitoring of workers exposed to solvents.

## 5. ACTUAL APPROACH FOR THE CLASSIFICATION OF CHEMICAL PREPARATIONS

Two other directives concern the classification and labelling of specific preparations: (i) concerning paints and varnishes and (ii) concerning pesticide preparations. At present no directive covering all the other chemical preparations exists. In order to achieve such a directive, the commission services, at the request and in close collaboration with the experts of Member States, have studied a method of evaluation of the toxicity of a chemical preparation. The method, that the commission services are discussing with the delegates of the Member States, consists of the following:

1. The characterisation of each constituent by an index: this index is a number representing the dilution degree able to render the substance not dangerous as defined in the Sixth Amendment, and furthermore, representing a numerical estimate of the more severe effect presented by the substance during acute toxicity, non-lethal effect after single and prolonged exposure. Particular relevance is given to the method of administration.

2. The estimation of the global toxicity of a preparation by applying a simple equation:

$$\sum (P \times I)$$

where $P$ is the percentage and $I$ is the index of each constituent.
3. The choice of the higher value obtained applying the same equation for the three methods of administration. In inhalatory administration, particular relevance is given to the value of the vapour pressure of each constituent by a simple modification of the original formula.
4. The classification of the preparation comparing the value obtained with the reference values within the three classes of toxicity, as referred to in the Sixth Amendment.

The other toxicological effects like irritation, corrosivity, sensitisation, mutagenicity, teratogenicity and carcinogenicity are considered in the classification on the basis of a system of limiting concentrations.

## REFERENCES

1. 67/548/CEE Directive du Conseil du 27 juin 1967 concernant le rapproche-ment des dispositions législatives, réglementaires et administratives relatives à la classification, l'emballage et l'étiquetage des substances dangereuses. *J. Off. Commun. Eur.*, 1967, **196**, 1.
2. 79/831/EEC Council Directive of 18 September 1979 amending for the sixth time Directive 67/548/EEC on the approximation of the laws, regulations and administrative provisions relating to the classification, packaging and labelling of dangerous substances. *Off. J. Eur. Commun.*, 1979, **259**, 10.
3. 73/173/EEC Council Directive of 4 June 1973 on the approximation of Member States' laws, regulations and administrative provisions relating to the classification, packaging and labelling of dangerous preparations (solvents). *Off. J. Eur. Commun.*, 1973, **189**, 7.
4. 80/781/CEE Directive du Conseil du 22 juillet 1980 modifiant la Directive 73/173 CEE concernant le rapprochement des dispositions législatives, réglementaires et administratives des Etats Membres, relatives à la classification, l'emballage et l'étiquetage des préparations dangereuses. *J. Off. Commun. Eur.*, 1980, **299**, 57.
5. Council Resolution of 29 June 1978 on an action programme of the European Communities on Safety and Health at Work. *Off. J. Eur. Commun.*, 1978, **165**, 1.
6. 80/1107/EEC Council Directive of 27 November 1980 on the protection of workers from the risks related to exposure to chemical, physical and biological agents at work. *Off. J. Eur. Commun.*, 1980, **327**, 8.

# 20

# FLAMMABILITY OF SOLVENTS*

F. C. Lloyd

*ICI Organics Division, Manchester, UK*

## ABSTRACT

*Despite the existence of legislation aimed at avoiding fires and explosions during the handling of solvents, they are involved in a large number of incidents each year. This chapter discusses the parameters used to characterise the flammability of solvents, and looks at their significance with regard to safe handling and use. Sources of energy that are capable of causing the ignition of flammable concentrations of solvent vapours are also considered.*

## 1. INTRODUCTION

Fires and explosions that involve flammable solvents are frequent occurrences in industry, and occasionally a major incident reaches the News programmes on TV and the front pages of our newspapers, and becomes the focus of attention.

Surprisingly, although the use of flammable solvents is extremely widespread, many users do not understand fully the nature of the materials with which they are working, nor the background to the legislation which controls the handling of some of these potentially hazardous substances.

Currently, the statutory instruments that cover the handling of flammable solvents are the Petroleum (Consolidation) Act, 1928[1] and the Highly Flammable Liquids and Liquefied Petroleum Gases

* Presented at the 3rd Solvents Symposium in 1980 and revised in 1983.

Regulations 1972.[2] The 1928 Act considers basically petroleum spirit, whilst the 1972 Act covers liquefied petroleum gases and all other liquids that can form flammable vapour atmospheres at temperatures below 32 °C, excluding those covered by the Petroleum (Consolidation) Act, 1928. The legislation is concerned with volatile solvents and so covers the most hazardous materials, but it has to be realised that materials that fall outside the Acts can be just as dangerous as those within them under some circumstances.

The following discussion will consider the flammability characteristics of solvents and deal briefly with sources of ignition.

## 2. FLAMMABILITY CHARACTERISTICS

When a liquid is placed in an open container it will evaporate away at a rate that depends, other things being equal, on the nature of the liquid. For instance, ether will evaporate more quickly than toluene, which in turn evaporates at a higher rate than, say, kerosene. As a liquid evaporates, mixtures of vapour and air are formed above it, and, if the material is oxidisable, these mixtures may be flammable. There is a range of parameters that is used to describe flammability, and these are considered in the following sections.

### 2.1. Limits of Flammability

Mixtures of solvent vapours in air (or oxygen) can only be flammable over a restricted range of concentrations, the limits of which are termed the lower and upper explosive limits. The lower explosive limit (LEL) is the minimum concentration for which flame propagation can occur once ignition has been effected, and the upper explosive limit (UEL) is the maximum concentration that will permit flame propagation following ignition.

Mixtures above the UEL are over-rich, containing an excess of fuel, and it should be appreciated when considering the flammability of a system that such mixtures will move readily into the flammable region if they are diluted with air. Measured limits depend on the direction of flame propagation and the dimensions of the vessel used in the test. Upward propagation in a vertical tube of at least 50 mm diameter gives the most appropriate conditions for the majority of mixtures at ambient temperatures and pressures.[3]

Considerable data on limits of flammability have been published by the US Bureau of Mines[3,4] from information collected over many years, and which is regarded as reliable. Limits for some common materials are shown in Table 1. Limits of flammability are influenced by temperature and pressure, the flammability range being widened by increases in either, as shown in Figs 1 and 2. Limits usually refer to normal flames initiated by high temperature sources (e.g. electric sparks), and involving little or no delay time between the application of the source and the appearance of the flame. Under certain conditions, however, cool flames occur which often increase the UEL quite dramatically into the region which previously was regarded as over-rich.[5] Such flames, which are only mildly exothermic, require a

TABLE 1
Limits of Flammability for the Vapours of Some Common
Materials[3]

| Material | Limits of flammability (volume per cent) | |
|---|---|---|
| | Lower | Upper |
| Acetic anhydride | 2·7 | 10 |
| Acetone | 2·6 | 13 |
| n-Amyl acetate | 1·0 | 7·1 |
| Aniline | 1·2 | 8·3 |
| Benzene | 1·3 | 7·9 |
| n-Butanol | 1·7 | 12 |
| n-Butyl benzene | 0·82 | 5·8 |
| Butyl Cellosolve | 1·1 | 11 |
| Cumene | 0·88 | 6·5 |
| Cyclohexane | 1·3 | 7·8 |
| Diethyl ether | 1·9 | 36 |
| Dimethylformamide | 1·8 | 14 |
| Dioxan | 2·0 | 22 |
| Ethanol | 3·3 | 19 |
| n-Heptane | 1·05 | 6·7 |
| Methanol | 6·7 | 36 |
| Methyl iso-butyl ketone | 1·2 | 8·0 |
| Methyl Cellosolve | 2·5 | 20 |
| Methyl ethyl ketone | 1·9 | 10 |
| iso-Propyl alcohol | 2·2 | 14 |
| Pyridine | 1·8 | 12 |
| Tetralin | 0·84 | 5·0 |
| Toluene | 1·2 | 7·1 |
| p-Xylene | 1·1 | 66 |

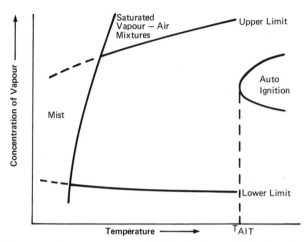

Fig. 1. Influence of temperature on limits of flammability in air at constant pressure.

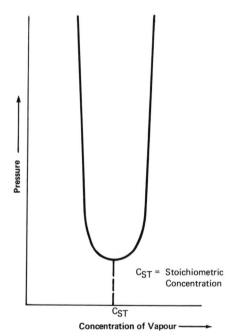

Fig. 2. Effect of pressure on limits of flammability at constant temperature.

controlled temperature, usually about 300–400 °C, applied for some
time for their initiation.

The normal flame range centres approximately at the stoichiometric
concentration (i.e. the theoretical mixture for complete combustion),
whilst the cool flame range centres on concentrations that are most
reactive in slow combustion.[5] These ranges may be separated by a
group of mixtures that will not propagate either type of flame.

At high pressures a two-stage ignition process induced by cool flames
occurs, and this leads to violent explosions in mixtures approaching the
stoichiometric value (Fig. 3).

Although a widening of limits occurs with cool flames, second-stage
ignitions resulting in explosive violence tend only to occur in the
flammable range for normal flames, so that in a safety context the
effects of cool flame phenomena on the limits of flammability do not
need to be considered. However, the initiation conditions should be

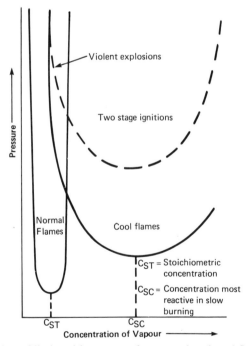

Fig. 3. Variation of limits with pressure for normal and cool flames at constant
temperature.

taken into account when examining the potential sources of ignition in a system.

The addition of increasing amounts of inert gases to a flammable mixture has the effect of narrowing the limits and eventually making the mixture non-flammable, which is the procedure adopted in inert gas blanketing. The lower limit is only slightly affected because at the lower limit the excess oxygen present is acting merely as a diluent, and replacing it with inert gas does not significantly affect the system. Nearer the upper limit, the inert gas replaces some of the fuel which is present in excess, thus depressing the limit. The efficiency of inert gases in inhibiting combustion is directly related to their molar heat capacity, and this is illustrated in Fig. 4 for methane,[4] which rates the gases $Ar < He < N_2 < H_2O < CO_2$.

In general terms, the limiting oxygen concentrations for non-flammable atmospheres with most hydrocarbon solvents are about 12%(v/v) using nitrogen and 14%(v/v) using carbon dioxide. In practice, to allow for variations in mixture composition, oxygen contents are reduced to well below these limiting values; 5–8% by volume being common.

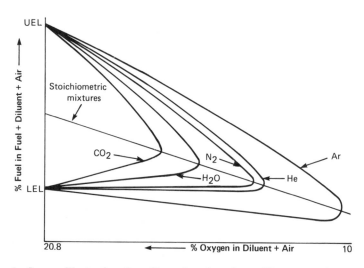

Fig. 4. Curves illustrating the effect of various inert diluents on the limits of flammability of a typical hydrocarbon.

## 2.2. Flash Point

Limits of flammability are not very convenient as working tools. It is frequently more helpful to know the conditions under which flammable mixtures of vapour in air will be formed, and the parameter most used in this context is the flash point. Flash point is that temperature to which a liquid must be heated before it will produce a flammable mixture of vapour in air that will ignite momentarily. At the flash point continuous combustion does not occur; this takes place at a higher temperature known as the fire point.

Flash point is influenced by a number of factors which are: equipment design, size of sample, ignition source, temperature control, ambient pressure, sample homogeneity, draughts and operator bias.[6] Over many years test equipment and methods have been set out in various national standards,[7,8] and the careful use of these will avoid variations due to most of these factors. Corrections do have to be made, however, for variations due to changes in ambient pressure. Flash points are normally corrected to standard barometric pressure (760 mm Hg); the correction usually being to add, or subtract, 0·03 °C to the measured value for each millimetre of mercury below or above 760 mm, respectively.[6]

The most extensively used pieces of equipment for determining flash points are the Abel Closed Cup (Fig. 5), Pensky Martens Closed Tester, the TAG Closed and Open Cups and the Cleveland Open Cup (Fig. 6).

Flash points determined in closed-cup equipment are normally several degrees below those obtained using an open cup because they are determined on a saturated vapour–air mixture, whereas in the open-cup test the vapour has free access to the air and will be slightly less concentrated at a given temperature.

Table 2 shows some data for common materials and compares closed- and open-cup values. The quoted flash points are for pure materials, and it should be appreciated that small amounts of more volatile contaminants can markedly affect the data. Extensive flash-point data are available in the literature,[9,10] but care should be taken to ensure that a particular result is applicable to the actual material being considered.

In flash-point testing, great care is needed to ensure that the volatiles are not driven off before the flame is applied, and that the atmosphere within the cup does not become inert. This can happen when examining materials such as halogenated hydrocarbons or aqueous solutions. In

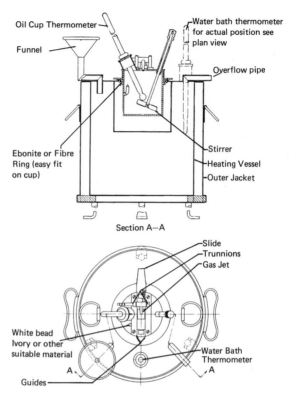

Fig. 5. Abel Flash-Point Apparatus.

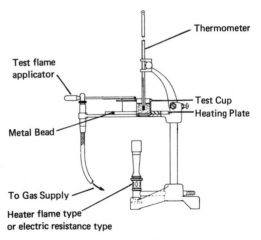

Fig. 6. Cleveland Open-Cup Apparatus.

TABLE 2
Flash-Point Data for a Range of Common Materials

| Material | Flash point (°C)[a] | |
|---|---|---|
| | Closed cup | Open cup |
| Acetanilide | 174 | 174 |
| Acetic acid (glacial) | 40 | 43 |
| Acetic anhydride | 49 | 54 |
| Acetone | −17 | −9 |
| n-Amyl acetate | 25 | 27 |
| iso-Amyl alcohol | 43 | 46 |
| Benzyl chloride | 60 | 74 |
| n-Butanol | 29 | 44 |
| iso-Butyl alcohol | 28 | 39 |
| 1,2-Diaminoethane | 34 | 37 |
| p-Dichlorobenzene | 66 | 74 |
| Dimethyl sulphate | 83 | 116 |
| Dioxan | 12 | 18 |
| Ethane diol | 111 | 116 |
| Ethyl acetate | −4 | −1 |
| Glycerol | 160 | 177 |
| Methanol | 10 | 16 |
| Methyl iso-butyl ketone | 17 | 24 |
| Phenol | 79 | 85 |
| iso-Propyl alcohol | 12 | 16 |
| Toluene | 4 | 7 |
| p-Xylene | 25 | 39 |

[a] These data may not have been obtained using the same test
procedure in each case.

these cases it may be necessary to deviate from the standard procedure
and blow out the cup with clean air between each ignition attempt.

Flash point can be regarded as a lower temperature limit of
flammability, and similarly an upper temperature limit can be defined;
in this way the temperature range over which a closed vessel will
contain a flammable atmosphere in the presence of excess liquid can be
obtained.

## 2.3. Auto-Ignition Temperature
Mixtures of solvent vapours in air or oxygen will ignite spontaneously
if the temperature is high enough. The lowest temperature at which
ignition will occur spontaneously is termed the auto-ignition tempera-

TABLE 3
Minimum Auto-Ignition Temperatures (AIT)
in Air at Atmospheric Pressure[3]

| Material | Minimum· AIT(°C) |
|---|---|
| Acetic acid (glacial) | 465 |
| Acetone | 465 |
| Benzene | 560 |
| Cyclohexane | 245 |
| Diethyl ether | 160 |
| Dimethyl formamide | 435 |
| Dioxan | 265 |
| Ethanol | 365 |
| n-Heptane | 215 |
| Methanol | 385 |
| Methyl Cellosolve | 380 |
| iso-Propyl alcohol | 440 |
| Toluene | 480 |
| p-Xylene | 530 |

ture. Unfortunately, auto-ignition temperatures are not absolute constants, but depend very much on the method by which they were obtained. The most widely used test method is that described in ASTM D2155-66 in which small quantities of liquid are injected into a 200 cm$^3$ conical flask which has been heated to a known temperature. The lowest temperature at which ignition occurs is designated the auto-ignition temperature.

Auto-ignition temperatures are significantly reduced by increases in pressure above atmospheric, and are further complicated by the possible presence of cool flames.[11] Cool flames normally occur in the 300–400 °C region, and so a search can be continued at higher values and, provided the cool and normal flame regimes are separated, the true value can be found. Where, however, the regions are not separated, it becomes necessary to base the determination on the appearance of the flame and this calls for considerable experience. Table 3 gives data obtained by Zabetakis at the US Bureau of Mines for a range of materials.

## 2.4. Mists

Thus far we have considered flammability as related to vapours, and suggested that flammable fuel–air mixtures will not occur at temperatures below the flash point. In the case of vapours this is true, but it is possible for fine droplet mists of solvents to be flammable well below the flash point,[12] and in solvent-handling systems this needs to be considered.

In practice, mists are normally generated by two systems: condensation and atomisation. In an 'uncontrolled' system, condensed mists are composed of droplets whose diameters do not exceed about 0·02 mm, whilst atomised sprays contain a few drops below 0·02 mm, but will include diameters up to about 1–2 mm.

If the liquid from which the mist is formed is combustible, the mist will be capable of propagating flame provided that the quantity of material present is enough. In a confined situation, therefore, explosions could occur, and this is true even if the temperature is too low for the liquid to contribute any significant amount of vapour to the system.

### 2.4.1. Limits of Flammability of Mists

Various workers[12–15] have shown that, as in the case of vapours, mists have lower and upper explosive limits. There is, however, little information on upper limits because of the extreme difficulty in carrying out reliable measurements due to the lack of uniformity of mists which it is virtually impossible to avoid at high concentrations.

Lower explosive limits are markedly affected by droplet size and experiments have shown[13] that for mists arising from the uncontrolled condensation of saturated vapour the droplet size is small and that the LEL, expressed as weight of liquid/unit volume of air, is approximately the same as for a vapour–air system. In such a system, the droplets (less than 0·014 mm diameter) have time to vaporise completely by heat transfer before they enter the flame front. Thus the flame is propagating into a vapour–air mixture so that the mist can be expected to behave in a similar manner to a vapour–air mixture. Larger droplets cannot vaporise completely before they enter the flame-front, and so the mechanism changes to the burning of a layer of vapour around individual droplets and not a continuous vapour mixture—flame propagation being by the transfer of combustion from the vapour layer of one drop to that of another.

The equipment used to determine the limits of flammability for mists is basically the same as that used for vapours, and when it is used for mists produced by atomisation, which have larger drop sizes than condensation mists, the results indicate a reduction in lower explosive limits as the drop size increases. In mists containing droplets whose diameters exceed about 0·02 mm, the drops may travel downwards faster than the air, and if this is so, the amount of liquid entering the flame-front relative to the air will be greater than when the drops and air travel at the same rate, as happens with smaller droplets.

This gives rise to the concept of kinetic concentration and Fig. 7 shows the correlation between the static concentration (weight of liquid per unit volume of air) and kinetic concentration (weight of liquid per unit volume of air entering the flame-front) at the lower explosive limit for tetralin–air suspensions.[12] The graph shows a much smaller fall in lower explosive limit with increasing drop size based on kinetic concentrations than for static concentrations. However, in most practical situations 'static' limits, which are easier to measure, are adequate for determining whether or not flammable conditions exist.

The introduction of a non-flammable gas to the air supporting a mist will affect the flammability of the mist in a manner that depends on the droplet size, the gas used and the amount introduced.

Burgoyne and Taylor obtained data for tetralin that showed that for droplet diameters of 0·01 mm and below, an addition of about 40% (v/v) nitrogen to the supporting atmosphere was necessary to avoid propagation.[12] This is close to the value for hydrocarbon vapours generally and is equivalent to a minimum oxygen concentration to support combustion of about 12% (v/v). With larger droplet sizes a

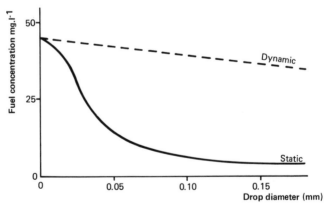

Fig. 7. Comparison of static and kinetic concentrations at the lower limit of flammability for a range of drop sizes.[13]

somewhat higher proportion of inert gas is needed. In practical terms this is not a serious change since a safety factor will normally be included in any system where the addition of inert gases is being used to suppress flammability. The value of 5–8% (v/v) quoted earlier for flammable vapour mixtures will be generally satisfactory for mists.

The flammability of mists and sprays is a topic on which research is currently being undertaken by various organisations, which are seeking detailed answers to problems such as: 'At what temperature below the flash point of a liquid does a mist become non-flammable? How is this related to droplet size? What influence does the energy and nature of the ignition source have on flammability?' and so on. What is clear, however, is that mists formed from droplets having very small diameters (less than 0·02 mm) behave essentially as if they were vapours.

## 3. SOURCES OF IGNITION

The sources of energy that are capable of causing the ignition of flammable solvent–air mixtures can be conveniently combined into four groups:

> flames and smouldering
> hot surfaces
> friction and impact
> electrical discharges

### 3.1. Flames and Smouldering
Under this heading come all naked flames: both established, such as matches, gas burners, pilot lights, welding torches, etc., and transient, as, for example, the flame emitted from the exhaust of an engine. Smouldering can be considered to cover any form of incandescent material. Typical sources of smouldering are pyrophoric deposits such as iron sulphide and certain catalysts, and, of course, cigarettes.

There are many ways in which flames and smouldering can occur in industrial plant, and they form probably the most potent source of ignition. No distinction is drawn between flames and smouldering as potential initiators because of the ease with which many smoulders can be caused to inflame. Flames or smouldering in the presence of a flammable solvent vapour or mist atmosphere will certainly cause ignition.

### 3.2. Hot Surfaces

Hot surfaces occur widely in industry and may cause ignition of flammable solvent–air mixtures either directly, or indirectly. Direct ignition will occur if the temperature of a surface is such that it can heat a flammable mixture to a temperature above its auto-ignition temperature, whilst indirect ignition is the result of the burning or smouldering of some material that has been initiated by the hot surface.

Typical hot surfaces are the walls of furnaces and ovens, surfaces of electrical equipment (e.g. motors, lamps, etc.), exhausts on diesel engines, heating pipes, etc. Also included are the hot areas produced by such operations as grinding, welding and cutting although in the latter two cases the presence of a flame will give a greater risk. Hot surfaces can be initiators for the cool flames referred to earlier in the discussion on limits of flammability, and should be taken into account when considering hot surfaces as sources of ignition.

### 3.3. Friction and Impact

Friction and impact can produce hot areas and/or incandescent sparks. Hot areas mostly arise from binding friction (e.g. the rubbing of a moving part on a stationary part), and as sources of ignition they are considered as hot surfaces.

The probability of ignition due to incandescent sparks arising from impacts depends largely on the impacting materials. The sparks produced during impacts involving aluminium, magnesium or their light alloys and rusty metal, titanium, and grit or rock are highly incendive, and will almost certainly ignite flammable solvent–air mixtures.

### 3.4. Electrical Discharges

Discharges can arise from two basic sources: electrical power and electrostatic charging. Power from the mains or batteries is a powerful source of sparks that have energies greater than that required to ignite flammable solvent vapour atmospheres. The sparks that can occur during the normal operation of commutator motors, circuit breakers, switches, fuses, and so on, can ignite flammable atmospheres as can those sparks arising from electrical failures, overloading in windings, damaged cables, etc.

Electrostatic charges are generated when insulating solvents are handled and processed, and it is the accumulation of these charges than can give rise to sparks that are capable of igniting flammable vapour atmospheres. Electrostatic spark discharges can occur from insulated conductors (e.g. plant items, drums, personnel), bulk liquid, droplet mists and insulating plastic materials. Considerable work has been done in the field of electrostatics[15–17] that enables the likely energy of sparks in a given situation to be estimated, and comparison of these estimates with the spark energies needed to cause ignition enables an assessment of the likelihood of ignition to be made.

Spark energies in the range 0·1–1·0 mJ are needed to ignite the vapours of hydrocarbon solvents in air at approximately the stoichiometric concentration. The energies of sparks available from insulated conductors depend on the electrical capacity of the conductor, its resistance to earth, and the charging conditions, and can be as high as several joules. The energies of discharges where these occur from bulk liquids, droplet mists and insulating plastic materials are usually limited by high resistances, but are capable of igniting some flammable solvent vapour–air mixtures especially close to the stoichiometric concentration.

Lightning is, of course, a special case of an electrostatic discharge and is certainly able to cause the ignition of a wide range of flammable solvent atmospheres.

There are few data available on the ability of electrical discharges to ignite flammable mists of solvents, although it is likely that for very fine mists (drops less than 0·02 mm diameter) the energies necessary to cause ignition will be very similar to those required for vapour–air mixtures.

## 4. CONCLUSIONS

We can draw the following practical conclusions:

1. The occurrence of flammable vapour atmospheres can be predicted from the flash point and limits of flammability.
2. The auto-ignition temperature is usable only as a very rough guide to the maximum temperature to which a flammable atmosphere can be exposed without self-ignition.
3. The influence of cool flames on the limits of flammability can be ignored from the safety point of view.
4. Droplet mists of solvents can be flammable at temperatures below

the flash point of the liquid, and at very small drop sizes they behave in a similar manner to vapours.

5. The kinetic concentration at the lower limit of flammability for mists is the same as that for vapours on a weight per unit volume basis.

6. There is a wide range of sources capable of igniting flammable mixtures of solvent vapours or mists in air.

## REFERENCES

1. Petroleum (Consolidation) Act 1928, HMSO, London.
2. Highly Flammable Liquids and Liquefied Petroleum Gases Regulations 1972, HMSO, London.
3. M. G. Zabetakis, Bulletin 627, US Bureau of Mines, 1965.
4. H. F. Coward and G. W. Jones, Bulletin 503, US Bureau of Mines, 1952.
5. P. F. Jessen, *Technical Data on Fuels* (7th edn), British National Committee, World Energy Conference, London, 1977.
6. R. C. Lance, A. J. Barnard and J. E. Hooyman, *J. Haz. Mat.*, 1979, **3,** 107–19.
7. BS 3442, British Standards Institution, London, 1979.
8. ASTM D56, American Society for Testing and Materials, Philadelphia, 1979.
9. N. I. Sax, *Dangerous Properties of Industrial Materials* (5th edn), Van Nostrand–Reinhold, New York, 1979.
10. NFPA, *Fire Protection Guide on Hazardous Materials* (7th edn), Boston, 1978.
11. A. Melvin, *Technical Data on Fuels* (7th edn), British National Committee, World Energy Conference, London, 1977.
12. J. H. Burgoyne and D. M. Newitt, *Trans. Inst. Marine Eng.*, 1955, **67,** 255.
13. J. H. Burgoyne, *Inst. Chem. Eng. Symp. Ser.*, 1963, No. 15.
14. S. J. Cook, C. F. Cullis and A. J. Good, *Combust. Flame*, 1977, **30,** 309–17.
15. J. H. Burgoyne and L. Cohen, *Proc. Roy. Soc.*, 1954, **A225,** 375.
16. A. Klinbenberg and J. L. Van der Minne, *Electrostatics in the Petroleum Industry,* Elsevier, Amsterdam, 1958.
17. F. C. Lloyd, in *Solvents—The Neglected Parameter* (ed. by G. J. Kakabadse), UMIST publication, 1977, pp. 115–26.

# INDEX

**RETURN TO:**     **CHEMISTRY LIBRARY**

100 Hildebrand Hall • 510-642-3753

| LOAN PERIOD | 1 | 2 _1 Month_ | 3 |
|---|---|---|---|
| 4 | | 5 | 6 |

**ALL BOOKS MAY BE RECALLED AFTER 7 DAYS.**

Renewals may be requested by phone or, using GLADIS,
type **inv** followed by your patron ID number.

**DUE AS STAMPED BELOW.**

| JAN 03 2005 | | |
|---|---|---|
| | | |
| MAR 3 0 | | |
| | | |
| | | |
| | | |
| | | |
| | | |
| | | |
| | | |

FORM NO. DD 10        UNIVERSITY OF CALIFORNIA, BERKELEY
3M 5-04                Berkeley, California 94720–6000